Lecture Notes in Artificial Intelligence   2007

Subseries of Lecture Notes in Computer Science
Edited by J. G. Carbonell and J. Siekmann

W0246258

Lecture Notes in Computer Science
Edited by G. Goos, J. Hartmanis and J. van Leeuwen

Springer
Berlin
Heidelberg
New York
Barcelona
Hong Kong
London
Milan
Paris
Singapore
Tokyo

John F. Roddick   Kathleen Hornsby (Eds.)

Temporal, Spatial, and Spatio-Temporal Data Mining

First International Workshop, TSDM 2000
Lyon, France, September 12, 2000
Revised Papers

 Springer

Series Editors

Jaime G. Carbonell, Carnegie Mellon University, Pittsburgh, PA, USA
Jörg Siekmann, University of Saarland, Saabrücken, Germany

Volume Editors

John F. Roddick
Flinders University of South Australia, School of Informatics and Engineering
P.O. Box 2100, Adelaide 5001, South Australia
E-mail: roddick@cs.flinders.edu.au

Kathleen Hornsby
University of Maine, National Center for Geographic Information and Analysis
Orono, Maine, ME 04469-5711, USA
E-mail: khornsby@spatial.maine.edu

Cataloging-in-Publication Data applied for

Die Deutsche Bibliothek - CIP-Einheitsaufnahme

Temporal, spatial, and spatio-temporal data mining : first
international workshop ; revised papers / TSDM 2000, Lyon, France,
September 12, 2000. John F. Roddick ; Kathleen Hornsby (ed.). – Berlin ;
Heidelberg ; New York ; Barcelona ; Hong Kong ; London ; Milan ;
Paris ; Singapore ; Tokyo : Springer, 2001
  (Lecture notes in computer science ; Vol. 2007 : Lecture notes in
  artificial intelligence)
  ISBN 3-540-41773-7

CR Subject Classification (1998): I.2, H.3, H.5, G.3, J.1, F.4.1, H.2.8

ISBN 3-540-41773-7 Springer-Verlag Berlin Heidelberg New York

Springer-Verlag Berlin Heidelberg New York
a member of BertelsmannSpringer Science+Business Media GmbH
http://www.springer.de
© Springer-Verlag Berlin Heidelberg 2001
Printed in Germany

Typesetting: Camera-ready by author, data conversion by Boller Mediendesign
Printed on acid-free paper   SPIN: 10782002   06/3142   5 4 3 2 1 0

Foreword

This volume contains updated versions of the ten papers presented at the First International Workshop on Temporal, Spatial and Spatio-Temporal Data Mining (TSDM 2000) held in conjunction with the 4th European Conference on Principles and Practice of Knowledge Discovery in Databases (PKDD 2000) in Lyons, France in September, 2000.

The aim of the workshop was to bring together experts in the analysis of temporal and spatial data mining and knowledge discovery in temporal, spatial or spatio-temporal database systems as well as knowledge engineers and domain experts from allied disciplines. The workshop focused on research and practice of knowledge discovery from datasets containing explicit or implicit temporal, spatial or spatio-temporal information.

The ten original papers in this volume represent those accepted by peer review following an international call for papers. All papers submitted were refereed by an international team of data mining researchers listed below. We would like to thank the team for their expert and useful help with this process. Following the workshop, authors were invited to amend their papers to enable the feedback received from the conference to be included in the final papers appearing in this volume. A workshop report was compiled by Kathleen Hornsby which also discusses the panel session that was held.

In late 1998, a bibliography of temporal, spatial and spatio-temporal data mining research papers was compiled by John Roddick and Myra Spiliopoulou and published in SigKDD Explorations. Following the dramatic increase in interest in this area over the past few years, it was decided that an update to this bibliography was warranted and should be included in this volume.

It is our contention that temporal, spatial and spatio-temporal data mining holds many promises. We hope this volume, as well as the TSDM Workshop series, is useful in adding to the body of knowledge in this area.

January 2001

John F. Roddick
Kathleen Hornsby

Organization

Workshop Chairs

John F. Roddick, Flinders University, Australia
Kathleen Hornsby, University of Maine, USA

Program Committee

Rakesh Agrawal, IBM Almaden Research Center, USA
Markus Breunig, Ludwig-Maximilians University, Germany
Max Egenhofer, University of Maine, USA
Martin Ester, Ludwig-Maximilians University, Germany
Marie-Christine Fauvet, Grenoble, France
Jiawei Han, Simon Fraser University, Canada
Christian Jensen, Aalborg University, Denmark
Heikki Mannila, Helsinki University, Finland
Harvey Miller, University of Utah, USA
Raymond Ng, University of British Columbia, Canada
Donna Peuquet, Pennsylvania State University, USA
Chris Rainsford, Defence Science and Technology Organisation, Australia
Hanan Samet, University of Maryland, USA
Stefano Spaccapietra, Swiss Federal Institute of Technology (EPFL), Switzerland
Myra Spiliopoulou, Humboldt University, Germany
Paul Stolorz, Jet Propulsion Laboratory, USA
Alex Tuzhilin, New York University, USA

Table of Contents

Workshop Report - International Workshop on Temporal, Spatial, and Spatio-temporal Data Mining - TSDM 2000

Kathleen Hornsby

National Centre for Geographic Information and Analysis,
University of Maine,
Orono, Maine, ME 04469-5711, USA.
khornsby@spatial.maine.edu

1 Introduction

The International Workshop on Temporal, Spatial, and Spatio-Temporal Data Mining (TSDM2000) was held on September 12, 2000 in Lyon, France as a pre-conference workshop of the Fourth European Conference on Principles and Practice of Knowledge Discovery in Databases (PKDD'2000). Workshop co-chairs were Kathleen Hornsby (University of Maine, USA) and John Roddick (Flinders University, Australia).

The aim of the workshop was to bring together experts in the analysis and mining of temporal and spatial data and in knowledge discovery in temporal, spatial or spatio-temporal database systems as well as knowledge engineers and domain experts from associated disciplines.

The Call for Papers solicited papers relating to:

- accommodating domain knowledge in the mining process,
- complexity, efficiency and scalability of temporal and spatial data mining algorithms,
- content-based search, retrieval, and discovery methods,
- data mining from GIS,
- KDD processes and frameworks specifically catering for temporal and spatial data mining,
- mining from geographic and geo-referenced data,
- representation of discovered knowledge,
- sequence mining,
- spatial clustering methods,
- spatial data mining,
- spatio-temporal data mining,
- temporal and spatial classification,
- temporal association rule mining, and
- uncertainty management in temporal and spatial data mining.

Many of these topics can be treated from either a spatial or temporal perspective and one of the workshop goals was to explore the connections or common

J.F. Roddick and K. Hornsby (Eds.): TSDM 2000, LNAI 2007, pp. 1–4, 2001.
© Springer-Verlag Berlin Heidelberg 2001

ground between spatial and temporal data mining by bringing together specialists from both fields. Spatial and temporal data mining are important subfields of data mining. As many traditional data mining techniques have yet to be developed for such complex types of data, they offer many challenges for researchers. We received 16 submissions to the workshop and, after review by the workshop program committee, accepted 10 papers for presentation and inclusion in the proceedings. The workshop program was comprised of paper presentations and a panel moderated by Kathleen Hornsby (University of Maine).

2   Workshop Papers

The papers were organized into three sessions. In the first session, Yingjui Li, Sean Wang, and Sushil Jajodia described work on developing algorithms that detect temporal patterns over multiple levels of temporal granularity. Ray Hickey and Michaela Black then discussed refined time stamps for concept drift detection. This is an important topic for large databases where some or all of the classification rules in use change as a function of time. In the third paper in this session, Bin Zhang, Meichun Hsu, and Umesh Dayal presented a center-based clustering algorithm, the K-Harmonic Means. This approach overcomes a well-known problem with typical K-Means clustering algorithms, namely, the sensitivity of such algorithms to how the centers are initialized.

The second session included a paper by Richard Povinelli on time series data mining as applied to financial time series with a view to characterizing and predicting complex, irregular, and even chaotic time series. Ruixin Yang, Kwang-Su Yang, Menas Kafatos, and X. Sean Wang then discussed a clustering technique based on histograms of earth science data where the histograms are clustered and used to index the database allowing statistical range queries to be performed. Vladimir Estivill-Castro and Michael Houle proposed a set of fast randomized algorithms for computing robust estimators of location important for spatial data mining on multivariate data. Finally, Thomas Bittner described a method for spatio-temporal data mining based on rough sets where spatio-temporal regions and the relations between such regions are approximated by rough sets.

In the third session, Karine Zeitouni, Laurent Yeh, and Marie-Aude Aufaure discussed the implementation of join indices as tools for spatial data mining. With the introduction of join indices, many spatial data mining problems can be reduced or transformed into relational data mining problems on which mining can be easily and efficiently performed. Howard Hamilton and Dee Jay Randall addressed data mining of temporal data using domain generalization graphs where a domain generalization graph based on calendar attributes serves as both a navigational aid to users as well as a guide to heuristic data mining procedures. For the final formal paper, Vladimir Estivill-Castro and Ickjai Lee presented AUTOCLUST+, an automatic clustering algorithm that treats the problem of obstacles in spatial clustering.

3 Panel Session

The final session of the workshop was a panel discussion with panelists Howard Hamilton (University of Regina, Canada), Robert Laurini (Claude Bernard University of Lyon, France), and Myra Spiliopolou (then with Humboldt University, now with Otto-von-Guericke University, Germany). Panelists were asked to discuss research challenges and open research questions for temporal, spatial, and spatio-temporal data mining.

Robert Laurini responded to these with several open research questions. Firstly, it would be challenging to prove Tobler's *First Law of Geography* that states that everything is related to everything else but near things are more related than distant things (Tobler, 1970, p. 236). Given this, what would be the Second Law? Secondly, assuming continuity rather than discreteness poses challenges for modeling. Our current attribute-based systems lend themselves to a discrete view of phenomena, but continuous phenomena require a different treatment. Laurini also introduced the notion of spatio-temporal differential equations that require a different kind of knowledge. With respect to data mining, Laurini proposed that clustering techniques need to be developed based on properties other than distance and that rely not only on visual clustering.

Howard Hamilton described what he thought the near future might be likely to offer with respect to descriptive data mining, predictive data mining, and constraint-based querying. Of these three topics, he proposes that the last one is going to be particularly rich where, for example, a problem domain involves parameters of space and time and one determines what instances match these parameters (e.g., is an object by a lake, near a road etc.). Hamilton pointed out that data currently resides in relational databases but in the future these databases will be designed to accommodate the needs and uses of handheld, mobile devices where, for example, users must try to coordinate calendars, schedules, etc. Data mining will be relevant for these scenarios. With mobile data sources, devices will capture the time when data is recorded, and when data gets changed, and we will need to distinguish these times. For these cases, perhaps bi-spatio-temporal databases will be required. Also, in certain contexts, it might be necessary to account, for example, for periodic events. In general, more research is necessary to look at the special techniques required for spatio-temporal data mining.

Myra Spiliopolou referred to the interesting properties of time, for example, the evolution of associations among objects and how it can be important to observe the temporal dimension of data mining results. She pointed out that although Euclidean space is important, it might not be appropriate for many data mining tasks. Spiliopolou asked how spatial mining could be applied to domains that are not perhaps, typical GIS domains, such as DNA sequences and web mining where temporal sequences that are not time series exist. She also remarked that there are numerous application domains including e-commerce and the web, where spatio-temporal data mining can provide solutions. It is important to have linkages between these different communities in order to see how spatial and temporal mining can provide solutions for these important and large areas. For some of the application areas, it may be necessary to introduce other techniques that combine with spatio-temporal data mining in order to get meaningful results.

4   References

Tobler, Waldo, 1970. A computer movie simulating urban growth in the Detroit region. Economic Geography, 46(2), 234-240.

Discovering Temporal Patterns in Multiple Granularities*

Yingjiu Li, X. Sean Wang, and Sushil Jajodia

Center for Secure Information Systems, George Mason University,
Fairfax VA 22030-4444, USA
{yli2,xywang,jajodia}@gmu.edu

Abstract. Many events repeat themselves as the time goes by. For example, an institute pays its employees on the first day of every month. However, events may not repeat with a constant span of time. In the payday example here, the span of time between each two consecutive paydays ranges between 28 and 31 days. As a result, regularity, or temporal pattern, has to be captured with a use of granularities (such as day, week, month, and year), oftentimes multiple granularities. This paper defines the above patterns, and proposes a number of pattern discovery algorithms. To focus on the basics, the paper assumes that a list of events with their timestamps is given, and the algorithms try to find patterns for the events. All of the algorithms repeat two possibly interleaving steps, with the first step generating possible (called candidate) patterns, and the second step verifying if candidate patterns satisfy some user-given requirements. The algorithms use pruning techniques to reduce the number of candidate patterns, and adopt a data structure to efficiently implement the second step. Experiments show that the pruning techniques and the data structure are quite effective.

1 Introduction

An important aspect of knowledge discovery regards efficient methods to find out how events repeat themselves as the time goes by. Research in this area has been fruitful, resulting in algorithms to discover frequent episodes [1] and cyclic associations [2], among other interesting phenomena. Unfortunately, missing from most research reported in the literature is the use of multiple time granularities (years, months, days and etc.) in such repeating patterns. Indeed, most human activities are closely related to time granularities. Events repeat, but they may not repeat with a constant span of time. Rather, they may repeat with a particular pattern in terms of multiple granularities. Furthermore, even if the event does repeat with a constant span of time, the use of granularity to represent the repetition pattern may be much easier to understand to a human

---

* Part of work was supported by NSF with the grant 9633541, and by ARO with the contract number 096314. Work of Wang was also partially supported by NSF with the career award 9875114.

J.F. Roddick and K. Hornsby (Eds.): TSDM 2000, LNAI 2007, pp. 5–19, 2001.

user. The question we ask in this paper is, given a set of events, how we can quickly discover their repetition pattern in terms of multiple granularities.

Consider the employee paydays at an institute. Suppose these paydays are the 1st and the 16th days of each month. In this example, the differences between two consecutive paydays range from 13 to 16 days. Hence, if only days are considered, there does not appear to be a strong regularity. However, if months are considered together with the days, the event repeats itself with a very clear regularity as given above. Consider another example where the office hours of Professor Smith are on every Monday. Again, if only days are considered, we may only be able to say that Professor Smith's office hours happen every 7th day. With the granularity week considered, we will be able to say that the office hours actually happen on every Monday (i.e., first day of week), which conveys more meaningful information to a human user.

This paper starts with the above observations and defines the concept of a simple calendar-based pattern. For example, $\langle year : 2000, month :*, day : 1 \rangle$ is a simple calendar-based pattern which intuitively represents the first day of every month in year 2000. Here we use the intuitive wildcard symbol '$*$' to denote "every". More complex patterns can be defined based on these simple ones. In general, a calendar-based pattern defines a set of time points. A calendar-based pattern discovery problem is to find, with a user-given set of granularities, all calendar-based patterns that match a given set of events with respect to the user-required minimum support and confidence rates. The support rate of a pattern is the percentage of the events, among all the given events, that fall into the time points given by the calendar-based pattern, and the confidence rate is the percentage of the time points, among all the time points given by the pattern, that contain the given events. Roughly speaking, these rates signify how "tight" the pattern fits the given events.

Obviously, a naive algorithm for such a discovery problem is to generate all possible patterns and then check each one to see if the required support and confidence rates are satisfied. However, this method does not always work well. In some cases, due to the large number of possible patterns, it is clearly an inefficient method. Optimization is necessary. This paper studies four algorithms with different optimization methods. All of the algorithms repeat two possibly interleaving steps. The first step generates possible patterns (called candidate patterns), and the second step checks the support and confidence rates of the candidate patterns. Different algorithms use different methods to generate candidate patterns and use different ways to take advantage of the previously checked patterns to prune out the next batch of candidate patterns. The major performance indicator is the number of candidate patterns generated and checked. In addition, all the algorithms use an efficient data structure to support the checking step.

In order to validate our algorithms and study their efficiency and scalability, we performed a series of experiments on a synthesized data set. The results indicate the situations where particular algorithms work well.

In summary, the contribution of the paper is three fold. First, we identified the importance of granularities in knowledge discovery, especially the repeating temporal pattern discovery. Second, we designed algorithms, especially pruning techniques, to efficiently perform the discovery task. Third, we performed experiments to study the algorithms.

The rest of the paper is organized as follows: In Section 2, the concept of a simple calendar-based pattern is defined and the problem of calendar-based pattern discovery is formulated. In Section 3, algorithms for discovering simple patterns are presented. In Section 4, experimental results are presented and analyzed. Section 5 discusses related work. Section 6 concludes the paper.

2   Problem Formulation

In this section, we formulate the calendar-based pattern discovery problem. We start with defining calendar schemata and simple calendar-based patterns.

A *calendar schema* is a relation schema (in the sense of the relational database model [3]) which has *fields* f_i and corresponding *domains* D_i $(1 \le i \le n)$:

$$R = (f_n: D_n, \ f_{n-1}: D_{n-1}, \ \ldots, \ f_1: D_1)$$

where each field is a *time granularity*[1], and each domain is the union of the wild card symbol * and a subset of integers. The domain D_i may be omitted in a calendar schema when it is obvious.

Each tuple $\langle d_n, d_{n-1}, \ldots, d_1 \rangle$ of the calendar schema, where each d_i is the wildcard symbol '*' or an integer in D_i, is called a *calendar expression*. The calendar expression $\langle d_n, d_{n-1}, \ldots, d_1 \rangle$ defines "time periods" that are intuitively described by: "the d_1th granule, of the granules of f_1 that are *covered by*[2] the d_2th granule, of the granules of f_2 that are covered by the d_3th granule,..., of the granules of f_n." In the above description, if d_i is the wild card * (instead of an integer), then the phrase "the d_ith granule" need to be replaced by the phrase "each of the granules" or "every granule". For example, for the calendar schema $(week, day, hour)$, the calendar expression $\langle *, 1, 10 \rangle$ means "the 10th hour (or 10am) on the first day (i.e., Monday) of every week". Similarly, $\langle 1, *, 10 \rangle$ represents the time periods "the 10th hour (or 10am) of every day of week 1".

We call a calendar expression with no wildcards a *basic time unit*. A basic time unit represents one specific granule of f_1. For example, $\langle 1, 1, 10 \rangle$ in the schema $(week, day, hour)$ represents a granule of *hour* that is the 10th hour of the first day of week 1.

---

[1] Some commonly used time granularities are *century, year, month, week, day, hour, minute, second.* More abstractly, a *time granularity* is a countable set of non-overlapping *granules* where each granule is a subset of a *time domain.* Users may refer to [4] for a formal treatment of calendars.

[2] Granule g_1 covered by granule g_2 means that the time domain of g_1 is a subset of the time domain of g_2. For example, a *day* is covered by the *month* which the day falls in.

For simplicity, we require that each granule of f_i in the calendar schema be covered by a unique granule of f_{i+1}, and each granule of f_{i+1} is a union of some granules of f_i. For example, $(month, day)$ is allowed since each day is covered by a unique month and each month is a union of some days. However, the schema $(year, month, week)$ is not allowed because not every week is covered by a unique month. It is often convenient and necessary for users to define their own granularities used in calendar schemata. For example, the 24 hours of a day can be partitioned into five parts, representing early morning, morning, work hour, night, and late night respectively, each part becoming a granule in the new granularity. The readers are referred to [4] for defining granularities.

With the notions of calendar schema R and calendar expression e defined above, we define *simple calendar-based patterns* that will be used in our pattern discovery.

A *simple calendar-based pattern* (SP) under a calendar schema $(f_n, f_{n-1}, \ldots, f_1)$ is a pair $\langle e, \Gamma \rangle$, where e is a calendar expression and $\Gamma = (\gamma_n, \gamma_{n-1}, \ldots, \gamma_1)$ is a *constraint*, and each γ_i is a set of granules of f_i. An SP defines "time periods" that are the same as that of the calendar expression e, except that the phrase "the d_ith granule" and "every granule" must be replaced by "the d_ith granule if it is in γ_i" and "every granule if it is in γ_i" respectively. For example, if calendar expression e is $\langle year : 1999, month : *, day : 16 \rangle$, and constraint Γ is $(all\text{-}years, all\text{-}months, business\text{-}days)$, then the SP has the semantics "the 16th days of each month of year 1999, if these days are business days."

Given a calendar schema R, a constraint Γ, and an input[3] Ω that is a multiset[4] of granules of f_1, we formulate the *calendar-based pattern discovery* as finding all SPs (e, Γ) where e is a calendar expression in R such that the *support* of each SP is greater than a user-given minimum support threshold and the *confidence* is greater than a user-given minimum confidence threshold. An SP has *support* s in the input Ω if $s\%$ of granules in Ω fall into the time periods given by SP. An SP has *confidence* c if $c\%$ of granules in the time periods given by SP contain some granule(s) of Ω. Roughly speaking, these rates signify how "tight" the SP fits the given input events. For example, given pattern $\langle *, 16 \rangle$ on the schema $(month, day)$ (and $\Gamma = (all\text{-}months, all\text{-}days)$) and a set of given paydays (as input), the support rate is the percentage of paydays (among all the paydays) that fall into the 16th days of each month, and the confidence is the percentage of the 16th days of each month that are paydays.

In reality, a lot of events in datasets (e.g., audit trail data) are stored together with their timestamps. The timestamps are usually represented in terms of multiple granularities (e.g., 1/15/99-14:41:44 is in terms of time granularities *month, day, year, hour, minute, second*). It is intuitively easy to convert such input (timestamps) to granules of f_1, we omit the details here.

---

[3] The input can be considered as a multiset of timestamps of interesting events.

[4] A multiset is an unordered collection of elements, like a set, but there may be several copies of each element in the multiset, and the number of copies is important. Note that sets cannot contain duplicate elements.

3 Algorithms

In this section, we will discuss four algorithms for solving the calendar-based discovery problem. All four algorithms use two interleaving steps. The first step generates possible (called *candidate*) patterns, and the second checks to see if a candidate pattern satisfies the given pattern requirements. The algorithms use different methods to generate candidate patterns in the first step but the same method for pattern checking in the second. The first step in three of the four algorithms interleaves with the second step.

To simplify our discussion, we initially assume that the input Ω is a multiset of basic time units under the calendar schema R. That is, the input can be writen as $\Omega = \{\langle \omega_n, \omega_{n-1}, \ldots, \omega_1 \rangle\}$, where each ω_i is an integer and the basic time unit $\langle \omega_n, \omega_{n-1}, \ldots, \omega_1 \rangle$ represents a granule of f_1. For example, if the calendar schema is $(month, day)$, we assume the input is like $\langle 1, 1 \rangle$ (January 1), $\langle 2, 16 \rangle$ (February 16) etc.. If the input is in other forms, we need to convert it first. For example, a timestamp "3am on Monday of the first week" can be converted to "early morning on Monday of the first week" if we stipulate that the granularity "early morning" be "0am to 6am of every day". The research on calendar algebra proposed in [5,6,7] can be used to realize such conversions. Due to space limitation, we do not address this issue.

3.1 Enumeration Algorithm (EA)

Given an input in the form of basic time units $\Omega = \{\langle \omega_n, \omega_{n-1}, \ldots, \omega_1 \rangle\}$, let D_i' for each $i = 1, 2, \ldots, n$ denote the union of set $\{*\}$ and the set of ω_i of all elements in the input. In other words, $D_i' = \{*, \omega_i | \langle \omega_n, \ldots, \omega_i, \ldots, \omega_1 \rangle$ in $\Omega\}$. Note $D_i' \subseteq D_i$ where D_i is the domain in the definition of the calendar schema. A natural idea for generating candidate SPs is to use each possible combination $\langle d_n, d_{n-1}, \ldots, d_1 \rangle$ where each d_i is taken from D_i', as a calendar expression to form the candidate SP $\langle e, \Gamma \rangle$.

For example, if we have input $\Omega = \{\langle 1995, 1, 16 \rangle, \langle 1999, 12, 1 \rangle\}$ for the calendar schema $(year, month, day)$, then $D_3' = \{1995, 1999, *\}$, $D_2' = \{1, 12, *\}$, $D_1' = \{1, 16, *\}$ and there are totally $|D_3'| \times |D_2'| \times |D_1'| = 27$ candidate patterns. Note that it is possible that some candidate patterns (e.g. the one with calendar expression $e = \langle 1999, 12, 16 \rangle$ in the above example) contain no granule in the input. We note that there is no need to check any other SPs not in the above set. Indeed, any SP with calendar expression $\langle d_n, \ldots, d_1 \rangle$ will have zero support and confidence rates if d_i is not in D_i' for some i since the corresponding time periods do not intersect the input data.

We call the algorithm which generates candidate patterns in this enumeration way the *Enumeration Algorithm (EA)*.

3.2 Enumeration Algorithm with Intra-level Pruning (EAP)

To optimize the process of generation of candidate SPs in EA, we have the following proposition:

---

(1) **initiate** Initialize two sets of calendar expressions: $S_1 = \emptyset$, $S_2 = \emptyset$, where
    S_1: calendar expressions whose corresponding SPs don't satisfy the support
    requirement;
    S_2: calendar expressions whose corresponding SPs satisfy both the support and
    confidence requirements.
(2) read in the input Ω, and construct a data structure CS. //see section 3.5 for
    detail.
(3) **for each** candidate SP with calendar expression e generated by EA, **do** the
    following
    (4) **if** there is e' in S_1 such that e is covered by e', **then** skip e,
    (5) **else** check pattern requirements (by using CS, see section 3.5) for the SP
        with e:
        (6) **if** the SP does not satisfy the support requirement, **then** insert e into
            S_1
        (7) **if** the SP satisfies the support and the confidence requirement, **then**
            insert e into S_2
(8) **output** S_2 which is the set of calendar expressions whose corresponding SPs
    satisfy the pattern requirements

Fig. 1. Algorithm EAP.

Proposition 1. *(Intra-level Pruning) If calendar expression e is covered by[5] calendar expression e', and if the SP with e' does not satisfy the support rate requirement, then the SP with e (with the same Γ) does not satisfy the support rate requirement.*

To see why this proposition holds, we only need to observe that the set of basic time units covered by e is a subset of the basic time units covered by e'. Therefore, the support rate for the SP with e is smaller than or equal to the support rate for the SP with e'. If the SP with e' does not satisfy the support rate requirement, then the SP with e does not either. For example, if the SP with the calendar expression $\langle year : 1995, month : 1, day : * \rangle$ does not satisfy the support rate requirement, then the SPs with calendar expressions $\langle 1995, 1, 1 \rangle$, $\langle 1995, 1, 2 \rangle, \ldots, \langle 1995, 1, 31 \rangle$ need not be verified against the pattern requirement.

By Proposition 1, we can prune out the SP with e if we already know that the SP with e' does not satisfy the support rate requirement and that e is covered by e'. This gives rise to the algorithm in figure 1, and we call this algorithm EAP. In Figure 1, the intra-level pruning is applied at line (4).

---

[5] Calendar expression $e = \langle d_n, \ldots, d_1 \rangle$ is covered by calendar expression $e' = \langle d'_n, \ldots, d'_1 \rangle$ if the time periods represented by e are covered by that represented by e'. Syntactically, e is covered by e' if for each $i = 1, \ldots, n$: either $d'_i = d_i$ or $d'_i =' *'$. For example $\langle 1999, *, 16 \rangle$ is covered by $\langle 1999, *, * \rangle$.

(1) **initiate** Initialize three sets of calendar expressions: $S_1 = \emptyset$, $S_2 = \emptyset$, $S_3 = \emptyset$, where

S_1: calendar expressions whose corresponding SPs don't satisfy the support requirement;

S_2: calendar expressions whose corresponding SPs satisfy both the support and confidence requirements;

S_3: calendar expressions whose corresponding SPs satisfy the support but not the confidence requirement.

(2) read in the input Ω, and construct a data structure CS. //see section 3.5 for detail.

(3) **for each** candidate SP with calendar expression e generated from a source point, **do** the following

    (4) **if** there is e' in S_1 such that e is covered by e', or there is a same e in S_2 or S_3, **then** skip e,

    (5) **else** check pattern requirements (by using CS, see section 3.5) for the SP with e:

        (6) **if** the SP does not satisfy the support requirement, **then** insert e into S_1

        (7) **if** the SP satisfies the support but not the confidence requirement, **then** insert e into S_3

        (8) **else** insert e into S_2

(9) **output** S_2 which is the set of calendar expressions whose corresponding SPs satisfy the pattern requirements

Fig. 2. Algorithm SPA.

3.3 Source Point Algorithm (SPA)

Another natural idea for generating candidate patterns is to generate possible calendar expressions from the elements in the input Ω. We will call each element in Ω a *pattern source point*. For each pattern source point $\langle \omega_n, \omega_{n-1}, \ldots, \omega_1 \rangle$, we use each possible combination $\langle d_n, d_{n-1}, \ldots, d_1 \rangle$ as a calendar expression to form a candidate SP, where each d_i is either ω_i or '$*$'.

For example, the candidate patterns for pattern source point $\langle 1995, 1, 16 \rangle$ are the ones with calendar expressions: $\langle *, *, * \rangle$, $\langle *, *, 16 \rangle$, $\langle *, 1, * \rangle$, $\langle 1995, *, * \rangle$, $\langle *, 1, 16 \rangle$, $\langle 1995, *, 16 \rangle$, $\langle 1995, 1, * \rangle$, $\langle 1995, 1, 16 \rangle$. Note that each candidate pattern contains at least one granule (the pattern source point) in the input, thus it may be better than EA in some cases (we will compare them in the experiment section).

Clearly, a pattern will contain no elements of the input (thereby having zero support and confidence) if it is not a candidate pattern given above, since any element in the input is contained in a candidate pattern generated from that element or pattern source point.

To optimize the process of generating candidate SPs, the duplicate calendar expressions generated by source points are eliminated before testing. For example, the calendar expression $\langle *, *, * \rangle$ will be generated from each pattern source point, however, the SP with this calendar expression is checked only one time. Furthermore, we also apply the intra-level pruning (proposition 1) as used in EAP. This gives rise to the algorithm in figure 2, and we call this algorithm the *Source Point Algorithm* (SPA). In Figure 2, line (4) applies the intra-level pruning by using S_1, and also eliminates duplicate patterns by using S_2 and S_3.

3.4   Granularity Based Algorithm (GBA)

The pattern discovery problem we discuss is for the calendar schema $R = (f_n, \ldots, f_1)$ and constraint $\Gamma = (\gamma_n, \ldots, \gamma_1)$. We can decompose this problem into a series of discovery problems. For each $i = n, \ldots, 1$, we define the discovery problem $\mathcal{P}_i$ the same way as in section 2 except using the calendar schema $(f_n, \ldots, f_i)$ and constraint $(\gamma_n, \ldots, \gamma_i)$. For each input element $\langle \omega_n, \ldots, \omega_1 \rangle$ in Ω, we get an element $\langle \omega_n, \ldots, \omega_i \rangle$ in the input to problem $\mathcal{P}_i$. Note the problem $\mathcal{P}_1$ is the problem we formulated before.

Clearly, each calendar expression $\langle d_n, d_{n-1}, \ldots, d_i \rangle$ of a candidate SP for problem $\mathcal{P}_i$ can be viewed as a concatenation of the calendar expression $\langle d_n, d_{n-1}, \ldots, d_{i+1} \rangle$ of a candidate SP for problem $\mathcal{P}_{i+1}$, with $d_i \in D'_i$ (see section 3.1 for definition of D'_i). Hence, each calendar expression for problem $\mathcal{P}_1$ can be generated from a calendar expression for problem $\mathcal{P}_2$ and D'_1. And each calendar expression for problem $\mathcal{P}_2$ can be generated from a calendar expression for problem $\mathcal{P}_3$ and D'_2, and so on. We have the following proposition for inter-level pruning.

Proposition 2. *(Inter-level Pruning) If the support rate of SP with the calendar expression $\langle d_n, d_{n-1}, \ldots, d_i \rangle$ and $\Gamma = (\gamma_n, \ldots, \gamma_i)$ is less than the support threshold in problem $\mathcal{P}_i$, then the support of SP with calendar expression $\langle d_n, d_{n-1}, \ldots, d_i, d_{i-1} \rangle$ and $\Gamma = (\gamma_n, \ldots, \gamma_i, \gamma_{i-1})$ is less than the support threshold as well in the problem $\mathcal{P}_{i-1}$ for each d_{i-1}.*

Again, the reason why the above proposition holds is because, intuitively, the set of basic time units covered by the calendar expression $\langle d_n, d_{n-1}, \ldots, d_i, d_{i-1} \rangle$ is a subset of the basic time units covered by the calendar expression $\langle d_n, d_{n-1}, \ldots, d_i \rangle$.

By Proposition 2, we may apply any algorithm we discussed earlier to each problem $\mathcal{P}_i$ for $i = n, \ldots, 1$ instead of applying the algorithm to $\mathcal{P}_1$ directly. For problem $\mathcal{P}_i$, we only check those calendar expressions which are concatenation of $d_i \in D'_i$ with some calendar expression of SP which satisfies the support requirement for problem $\mathcal{P}_{i-1}$. For example, if the support of SP with calendar expression $\langle *, 1 \rangle$ is less than sup_0 for problem $\mathcal{P}_2$, then all the SPs with calendar expressions $\langle *, 1, * \rangle$, $\langle *, 1, 1 \rangle$, $\langle *, 1, 2 \rangle, \ldots, \langle *, 1, 31 \rangle$ need not be checked for problem $\mathcal{P}_1$.

We call the above method the *Granularity Based Algorithm* (GBA). For simplicity we use EA in each problem $\mathcal{P}_i$ in this paper.

3.5   Pattern Checking

So far we have discussed three algorithms to generate candidate patterns. In each algorithm, we need to check whether each candidate pattern satisfies the suport and confidence requirements. A naive method to compute the support and confidence rates of an SP is to scan the entire input (which may be very large) each time we check an SP. Actually we only need a very small part of the input to check an SP if the calendar expression of the SP consists of not all symbol '*'. In order to quickly locate the part of the input for a candidate SP without scanning the entire input again and again, we use a data structure, called *calendar structure*, to check the pattern requirements more efficiently. The calendar structure allows a single pass over the input data to be sufficient. To exchange for speed, this method takes some extra memory to store the calendar structure.

A *calendar structure* (CS) for the calendar schema $R = (f_n, f_{n-1}, \ldots, f_1)$ is a *rooted tree* having the following components and structural properties (figure 3(a)): (1) There are n arrays (levels) of cells and each cell is associated with one node. Where the root node represents granularity f_n, the child nodes of the root cells represent $f_{n-1}, \ldots$, the leaf nodes represent f_1. (2) Each node is associated with an interval of cells at the same level. Each cell at a certain level represents a granule of that level (granularity). A cell's child node holds exactly all the cells whose corresponding granules are covered by the granule represented by the cell. (3) All leaf nodes are at the same level. Each leaf node must have at least one cell representing a granule in the input. (4) Each cell of a inner node has at most one child node. Each inner node as a whole must have at least one child node.

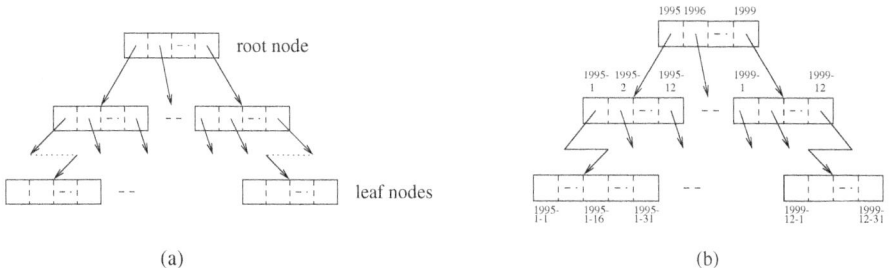

(a)   (b)

Fig. 3. (a) Calendar structure; (b) example of CS.

Figure 3 shows the calendar structure and an example of CS where the given calendar schema is $(year, month, day)$ and the input falls between 1995-1-1 and 1999-12-31.

By using the CS, it is easy to map a basic time unit (e.g. $\langle 1999, 12, 1 \rangle$ in Figure 3(b)) to a leaf cell of CS. Therefore, we scan the entire input one time, mapping each element (basic time unit) to a leaf cell of CS, and maintaining a counter for each leaf cell to record the number of elements that fall into the cell[6]. At the same time, we also get each D_i' (see section 3.1) that is used in different algorithms.

According to the definition of SP (see section 2), it is also easy to map a SP to a set of leaf cells of CS. For the example shown in Figure 3(b), to map an SP with $e = \langle 1999, *, 1\rangle$, $\Gamma = (all\text{-}years, all\text{-}months, business\text{-}days)$, to a set of leaf cells, we first select the root cell labeled 1999, then select all the cells that are in the child node of the root cell. Finally we select the first cells (if they represent business days) in each child note of the selected cells that are in the middle layer.

The support of an SP is simply the sum of counts in the mapped leaf cells of the SP divided by the total elements in the input. The confidence is the number of the mapped leaf cells of the SP that have counts greater than zero divided by the number of the mapped leaf cells. Clearly, the support rate and confidence rate computed this way are consistent with those defined in section 2.

4   Experiments

To study the performance of our algorithms, all four algorithms discussed in this paper, namely, EA, EAP, SPA, and GBA are implemented in the programming language Java and tested on a SUN UltraSPARC 200MHz workstation with 128MB memory. The basic parameters used in the experiments are summarized as follows:

– Different calendar schemata: $R_1 = (year, month)$, $R_2 = (year, month, day)$, $R_3 = (year, month, day, hour)$. The constraint for R_1 is $(1995\text{-}1999, all\text{-}months)$. The constraint for R_2 is $(1995\text{-}1999, all\text{-}months, business\text{-}days)$, and the constraint for R_3 is $(1995\text{-}1999, all\text{-}months, business\text{-}days, all\text{-}hours)$.
– Different inputs: The input is a synthetic multiset of basic time units $\langle \omega_n, \ldots, \omega_1\rangle$ for calendar schema $R = (f_n, \ldots, f_1)$, where ω_i is a random integer independently generated and uniformly distributed in its meaningful domain (D_i). Different inputs with sizes from 1 event to 100 million events are tested. The time of inputs spans 5 years for different calendar schemata. The maximum input is 600 Megabytes.
– Different support and confidence thresholds: different thresholds from 0.25% to 100% are used.

Figure 4(a) shows the execution time of different algorithms for pattern generation and check in relevance to the input size. We use the calendar schema R_2 and set the threshold for both support and confidence 0.75%. Figure 4(b) shows the time for reading in the input and constructing the CS, and this time is the same for different algorithms.

The curve in figure 4(b) shows that the scalability of each algorithm for reading in the input and constructing the CS is linear. And the curves in figure

---

[6] For GBA, we also maintain a counter for each cell at higher levels rather than just at the leaf level.

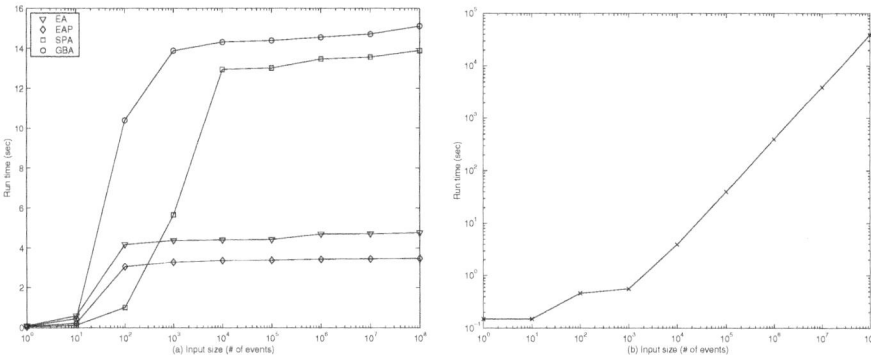

Fig. 4. Test result at different size of input for calendar schema R_2 and threshold (support and confidence) 0.75%. (a) Execution time of different algorithms for pattern generation and check. (b) Time for reading in the input and constructing the CS.

4(a) show that the scalability of each algorithm for pattern generation and check is better than linear on the input size.

From figure 4(a), we find that all the algorithms are insensitive to the input size after a threshold. The reason is as follows. As the input becomes large enough (after a threshold), almost every leaf cell in CS is mapped to from some input points, therefore the number of candidate patterns generated by each algorithm becomes stable, and insensitive to the input size.

EAP performs best in Figure 4 except that SPA is better for some very small inputs. The reason behind this is that the intra-level pruning works well for EAP. Although we apply intra-level pruning in SPA, the duplicate removal overweights the pruning benefits for large size of inputs. SPA performs best for some small inputs since the duplicate removal is not a main factor in these cases. We also know the inter-level pruning for GBA has more overhead than benefits in this figure.

Figure 5(a) shows the influence of the threshold for support and confidence to the performance of different algorithms, where the input size is 10^6 and the calendar schema is R_2. Clearly, EAP works best for most cases. Among the other three algorithms, SPA and GBA are better than EA for some small size of inputs, and this order is reverse for large size of inputs. Figure 5(c) shows the number of candidate patterns generated by different algorithms versus the number of patterns (denoted as SUC in Figure 5(c)) that satisfy the support and confidence requirement. We see that the intra-level pruning effect of EAP is the same as the SPA (but EAP saves the process of elimination of duplicate patterns), and the inter-level pruning for GBA works better for small inputs than for large inputs.

Note the threshold has no relationship with the time for reading in the input and constructing the CS, thus the CS curve is flat in Figure 5(b).

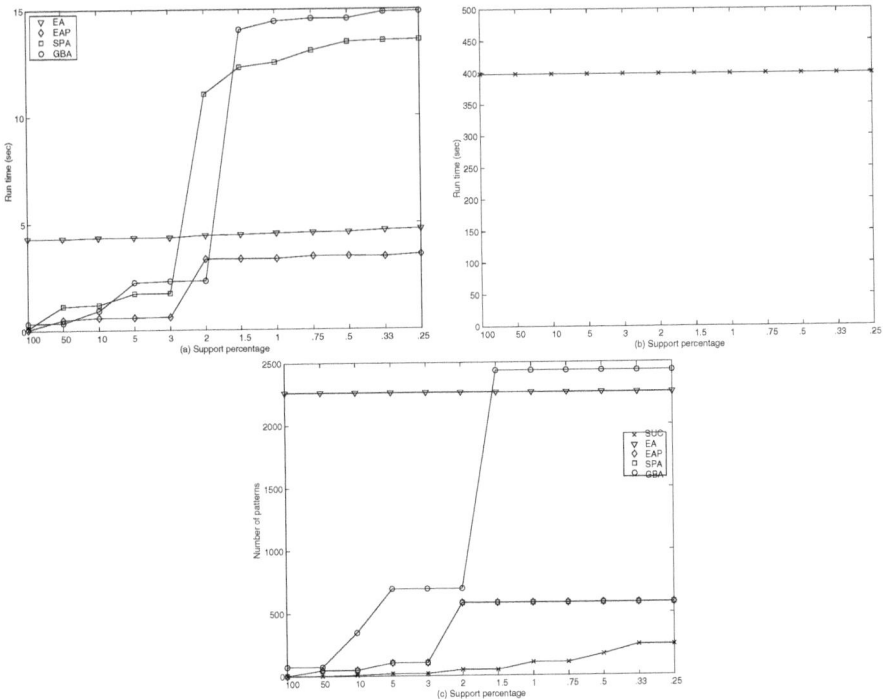

Fig. 5. Test result at various levels of threshold (support and confidence) for calendar schema R_2 and input size 10^6 events. (a) Execution time of different algorithms for pattern generation and check. (b) Time for reading in the input and constructing the CS. (c) Number of candidate patterns of different algorithms and number of successful patterns (SUC). Note the curves for EAP and SPA coincide.

We test the influence of calendar schemata to the performance of different algorithms in Figure 6, where the input size is 10^6 and the threshold for both support and confidence is 0.75%. Note we use logarithmic scale for the run time axis in Figure 6. If we consider the pattern space (i.e. the number of possible calendar-based patterns) is 78 for R_1, 2496 for R_2, and 62400 for R_3, the performance of different algorithms in Figure 6 has linear scalability to the pattern space (since in point of view of the pattern space, the x-axis in Figure 6 is also in logarithmic scale). We also note that for the large pattern space of R_3, GBA performs better than EAP, while in other cases, EAP is the best.

The effect of intra-level pruning of GBA is distinct for calendar schema R_3. In Figure 7, where the input size is 10^6 and the calendar schema is R_3, GBA is best for most thresholds. However, the performance of EAP is closed to that of GBA.

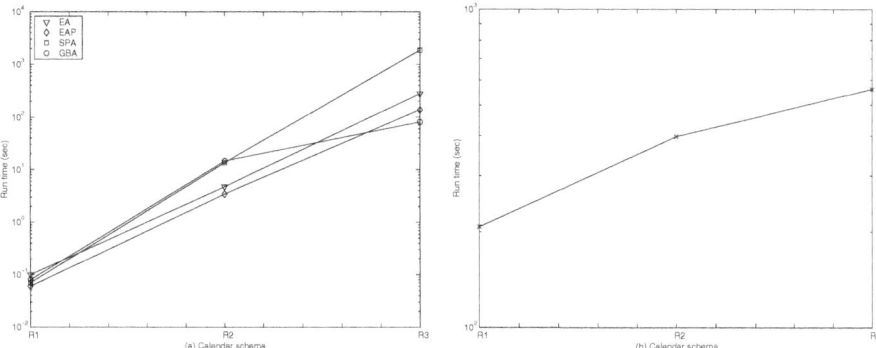

Fig. 6. Test result for different calendar schemata for the input size 10^6 events and threshold (support and confidence) 0.75%. (a) Execution time of different algorithms for pattern generation and check. (b) Time for reading in the input and construting the CS.

In summary, there are two methods to generate candidate patterns, one is enumeration method using D_i' as in EA, EAP, and GBA; another is pattern source point method as in SPA. The problem with the pattern source point method is to generate many duplicate patterns that need to be eliminated during pattern checking process. The time consuming process of duplicate removal undermines the pruning benefits in most cases except for some very small inputs.

There are also two pruning techniques, i.e. intra-level pruning and inter-level pruning. Intra-level pruning technique without duplicate removal (as in EAP) works best in most cases, while inter-level pruning is better for some calendar schemata that have larger pattern space.

It is interesting to know whether the benefits will overweight the overhead if we conbine the intra-level pruning with the inter-level pruning in some way. However, we do not address this issue in this paper.

5   Related Work

In general, our work is based on the basic research on the time granularities and the manipulation of time granularities (e.g., see [4,8,9]). It is also based on the general data mining and knowledge discovery research that have appeared in the literature. More specifically, this work is related to temporal data mining.

In temporal data mining, there are two basic directions. One direction is to use the temporal information to filter out irrelevant events, and only concentrate on events that happen at specific times. For example, [2,10,11] deal with this aspect of the knowledge discovery. In this direction, the ability to generate interesting patterns of time points is important. However, the user needs to provide such patterns. Temporal patterns are generally not discovered from the given data.

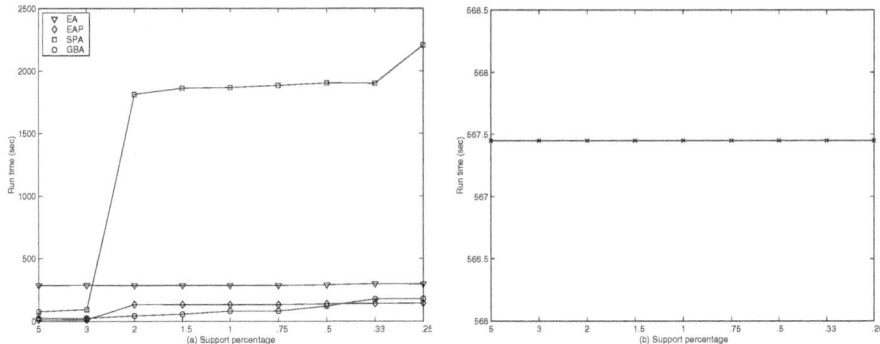

Fig. 7. Test result at various levels of threshold (support and confidence) for calendar schema R_3 and input size 10^6 events. (a) Execution time of different algorithms for pattern generation and check. (b) Time for reading in the input and construting the CS.

The other direction is to find the temporal patterns from given events. This paper belongs to this direction. This paper is unique since it is geared towards the discovery of temporal patterns in multiple granularities. We believe that the use of multiple granularities give rise many interesting patterns to be found.

Although we use the similar concepts of support and confidence, the issue addressed in this paper is different from discovery of frequent itemsets [12,13] or sequential patterns (possibly with regular expression constraints [14]). We are dealing with a multiset of time points rather than a multiset of itemsets. The relationship among time points is different from that among itemsets, and such relationship plays an important role in developing algorithms in this paper. In a broad sense, the work presented here and those of discovery of frequent itemsets and sequential patterns are complementary to each other. And an interesting question is how to combine them.

Research exists that deals with a more general discovery problem which looks for interesting events as well as the temporal patterns of these events at the same time. Examples towards such a general discovery problem include the papers [1,2], which provide methods to find frequent episodes and cyclic association rules, respectively. However, these papers do not deal with multiple granularities. Regarding this, we will address a particular issue, i.e., discovering temporal association rules in multiple time granularities, in a forthcoming paper.

6   Conclusion

This paper discussed ideas in what we believe are the first attempt to date to accommodate multiple granularities in discoverying temporal patterns. We first introduced the notions of calendar schemata, calendar expressions, simple calendar-based patterns and formulated the calendar-based pattern discovery problem. We then developed a group of algorithms and some efficient pruning techniques to solve the problem. We reported some experiment results, which shed some lights on the different performance characteristics of the different algorithms.

As mentioned in the related work section, the algorithms of this paper discovers temporal patterns under the assumption that the events are given. A future work direction will be to extend the work to methods that look for interesting events and their patterns at the same time. Here, the interestingness of the events may very well depend on the temporal patterns found for them.

References

1. Mannila, H., Toivonen, H., Verkamo, A. I.: Discovering frequent episodes in sequences (extended abstract). The 1st Conference on Knowledge Discovery and Data Mining. (1995) 210-215
2. Özden, B., Ramaswamy, S., Silberschtaz, A.: Cyclic association rules. Proceedings of the 14th International Conference on Data Engineering (1998) 412-421
3. Ramakrishnan, R.: Database management systems. McGraw-Hill. (1997) 21-37.
4. Bettini, C., Jajodia, S., Wang, X. S.: Time granularities in databases, data Mining, and temporal reasoning. Springer-Verlag, Berlin (2000)
5. Peng, N., Wang, X.S., Jajodia, S.: An algebraic representation of calendars (extended abstract). AAAI-2000 Workshop on Spatial and Temporal Granularity. Austin, Texas (2000)
6. Ramaswamy, S., Mahajan, S., Silberschatz, A.: On the discovery of interesting patterns in association rules. Proceedings of the 24th VLDB Conference. (1998) 368-379
7. Allen, J.F.: Maintaining knowledge about temporal intervals. Readings in Knowledge Representation. Morgan-Kaufman Publishers, Inc. (1985) 509-521
8. Bettini, C., Wang, X.S., Jajodia, S.: A general framework for time granularity and its application to temporal reasoning. Annals of Mathematics and Artificial Intelligence. **22** (1998) 29-58
9. Chandra, R., Segev, A., Stonebraker, M.: Implementing calendars and temporal rules in next generation databases. Proceedings of International Conference on Data Engineering. (1994) 264-273
10. Ramaswamy, S., Mahajan, S., Silberschatz, A.: On the discovery of interesting patterns in association rules. Proceedings of 24th VLDB Conference. (1998) 368-379
11. Bettini, C., Wang, X.Y., Jajodia, S.: Temporal semantic assumptions and their use in databases. IEEE Transactions on Knowledge and Data Engineering. **10** (1998) 277-296
12. Agrawal, R., Imielinski, T., Swami, A.: Mining association rules between sets of items in large databases. Proceedings of 1993 ACM-SIGMOD Conference. (1993) 207-216
13. Agrawal, R., Srikant, R.: Fast algorithms for mining association rules. Proceedings of the 20th VLDB Conference. (1994) 487-499
14. Garofalakis, M.N., Rastogi, R., Shim, K.: Mining sequential patterns with regular expression constraints. Proceedings of the 25th VLDB Conference. (1999) 223-234

Refined Time Stamps for Concept Drift Detection During Mining for Classification Rules

Ray J. Hickey and Michaela M. Black

School of Information and Software Engineering
Faculty of Informatics
University of Ulster
Coleraine, N. Ireland
BT52 1SA
{rj.hickey,mm.black}@ulst.ac.uk

Abstract. In many application areas where databases are mined for classification rules, the latter may be subject to *concept drift*, that is change over time. Mining without taking this into account can result in severe degradation of the acquired classifier's performance. This is especially the case when mining is conducted incrementally to maintain knowledge used by an on-line system. The TSAR methodology detects and copes with drift in such situations through the use of a time stamp attribute, applied to incoming batches of data, as an integral part of the mining process. Here we extend the use of TSAR by employing more refined time stamps: first to individual batches, then to individual examples within a batch. We develop two new decision tree based TSAR algorithms, CD4 and CD5 and compare these to our original TSAR algorithm CD3.

1 Introduction

In many application areas where large databases are being continually mined for classification rules, it is reasonable to suppose that the underlying rules will be subject to *concept drift* [1]. By concept drift we mean that some or all of the rules defining classes change as a function of time.

An example of a domain where this is likely to be true is fraud detection for, say, credit card or mobile phone usage. Here change may be prompted by advances in technology that make new forms of fraud possible or may be the result of fraudsters altering their behaviour to avoid detection. Mining in connection with personalisation of web sites provides a further example. Changes in topology of the web site may result in different user navigation behaviour, such as whether the appropriate submit button is reached quickly.

To see how drift can affect rules, consider the propensity of a customer to purchase a newly developed product. This may change substantially over time with innovative buyers who purchase early on in the lifetime of a product giving way to more conservative customers who hold back until the product has matured. Suppose the purpose of mining is to acquire rules for three classes identifying customer types for marketing purposes:

J.F. Roddick and K. Hornsby (Eds.): TSDM 2000, LNAI 2007, pp. 20-30, 2001.

class 1: likely to purchase within the next month
class 2: likely to purchase within the next one to three months
class 3: not likely to purchase within three months

Because, in a real world domain such as this, there is bound to be noise, we can represent rules using class probability distributions as in the following case:

```
if
      customer_status = regular & buyer_type = innovative
      & product_enquiry = yes  &  sex = male  &  age < 35
      & employed = yes
then
      (0.6, 0.1, 0.3)
```

where 0.6, 0.1 and 0.3 are the probabilities respectively of classes 1, 2 and 3 above.

Drift will alter this class distribution. Consequently, either the majority class will remain the same as in, for example,

$$(0.6, 0.1, 0.3) \rightarrow (0.7, 0.1, 0.2)$$

or a new majority class will emerge as in

$$(0.6, 0.1, 0.3) \rightarrow (0.4, 0.1, 0.5)$$

Both of these instances of drift have consequences for mining with the latter having greater impact.

Drift in rules may affect only part of a domain. For example in case of fraud, there may be different sources or types of fraudulent activity some of which remain more or less static over time whilst others are highly dynamic.

In addition to drift in rules there may be *population drift* [2]. That is the distribution of the description attributes used in mining may alter across the population. For example in the rule identifying customer type above, the percentage of people unemployed may change over time. Such changes need not affect the validity of the underlying classification rules as the latter are conditional structures, but in some domains there may be a correlation between the two types of drift. We are concerned here only with drift in rules and, for simplicity, will assume that there is no population drift.

Drift may happen suddenly, referred to as *revolutionary*, or may happen gradually over an extended time period, referred to as *evolutionary*; see [1] and [3]. In the former case we refer to the time at which drift occurred as the *drift point*. We can regard evolutionary drift as involving a series of separate drift points with very small drift occurring at each one.

The extent of prior knowledge about when drift is likely to occur depends very much on the application. For example in e-commerce, a substantial change to a company's web site may provoke changes in customer browsing behaviour and this can obviously be anticipated. The drift point will probably occur shortly after the redesign of the site (and is, perhaps, more likely to be revolutionary). Other easy to anticipate occurrences of drift are provided by seasonal factors such as holiday periods. Drift points may become apparent retrospectively as with interest rate changes and their affect on financial transactions such as house purchase. In contrast there will be domains in which it is not clear, even retrospectively, when drift might be present.

In spite of the above arguments, there is, in much of the current work in data mining, a tacit assumption that knowledge being mined for is static. Yet often the data source for mining has been collected over a considerable period of time making it more probable that drift has occurred on some occasion(s). In Machine learning, there has been work for more than a decade on concept drift for classification; see, for example [4], [5], [6] and [7]. Even there, though, the limited extent of the work is surprising given that human learning is quintessentially about adapting and updating knowledge in the light of change. There has also been work on time dependency for association rule mining; see, for example, [8] and [9].

The consequences of ignoring concept drift when mining for classification rules can be catastrophic [1]. Suppose that in a set of data collected over a period of time and prepared for mining, there has been a single revolutionary drift point that occurred at time point, T. Assume that the proportion of the data set collected before T was p_1 and that after T was p_2 (where $p_1 + p_2 = 1$). Suppose that before T the true underlying rule set was R_1 and that after T this became R_2. If the training data does not identify the time at which each example was collected then it is as if the underlying rule set throughout the collection phase was

$$R = p_1 R_1 + p_2 R_2$$

i.e. R is a mixture of R_1 and R_2. R is obtained by mixing individual rule class probability distributions in the proportions p_1 and p_2. (Before doing this, the rule sets must (if necessary) be brought to a 'common denominator', that is each must be refined so that their sets of rule antecedents are isomorphic.) The resulting mixed distributions will often have higher entropy than the individual distributions taking part in the mixture, assuming the latter are different. This is particularly so if rules before and after drift identify different majority classes. Thus the rules in R will tend to be noisier and, as a consequence will be more difficult to mine accurately.

The real drawback occurs when the mined rules are applied to classify unseen cases in the future. Assuming there has been no further drift after time point T, then any new case should be handled as an instance of a rule in R_2. It will, however, be treated as if it had been generated by a rule in R and classified accordingly. In [1] we refer to this activity as *cross-classification*. The success rate in classifying new examples on the basis that they were mined from R is the *cross-classification rate*. Depending on the extent of the differences between R_1 and R_2 the cross-classification rate can be quite low - much lower than it would have been if the data used for mining had been collected after time point T.

2   Drift and On-Line Mining: The TSAR Approach

The conclusions above apply not only to the case of one-off mining where drift might have occurred but are even more relevant for continual mining as part of the maintenance and improvement of a deployed classification system. Such a system might support the activities illustrated above: fraud detection, web site personalisation or marketing of new products. For these systems we require a regime of incremental mining in which the knowledge obtained from previous episodes of mining are regularly updated. An architecture for large-scale incremental learning appropriate for such mining was proposed in [1] where we argued that the natural *unit of incrementation* is a batch of data. That is, a new round mining is invoked automatically on receipt of a new batch. This produces an update of the knowledge used in the on-line performance task, e.g. vetting credit card transactions.

A major practical issue for incremental mining is the formation and size of batches. It may be that batches occur naturally, e.g. if they are delivered on a weekly or monthly basis. Alternatively it may be possible to request new data at any time. Intuitively, it is better for a batch to span as short a time period as possible to reduce the likelihood of drift having occurred during collection. Obviously the notion of a 'short time' is relative to the domain in question. It is also desirable to have batches be as large as possible to facilitate the mining process. In some applications it may be necessary to trade off batch size against length of the collection time period.

We argued above that mining without information about the time origin of data could lead to problems. The solution we proposed in [1] was to augment the data records with a time stamp and to actively use this in the mining process as a description attribute. Drift is then indicated by the time stamp attribute becoming a *relevant* attribute in the knowledge structure acquired during mining. We refer to this principle as *TSAR* (Time Stamp Attribute Relevance). Here we refer to any mining procedure that seeks to detect concept drift by this means as a TSAR procedure or algorithm. The simplest way to create a such a procedure for mining is to augment an existing classification learning algorithm, referred to as the *base learner*.

Before a TSAR procedure can be applied, the granularity of the time stamp must be decided upon. In [1] we considered a time stamp associated with a new batch, i.e. within the batch all time stamps were identical. We proposed an incremental concept drift detecting learning system, CD3 (CD =concept drift) which used ID3 and together with a post-pruning algorithm as its base learner (see figure 1). Only time stamps 'current' and 'new' were used. CD3 maintains a database of valid examples from the totality of data presented to the system in its lifetime. Each example in this current example set carries the time stamp value 'current'. When a new batch of data arrives, its examples are all time stamped 'new'. CD3 then learns classification rules from the combined current example set and the new batch using ID3 together with post pruning. Because the time stamp identifies a batch the problems due to mixing that were described above to not occur provided the drift occurs between batches. Even if drift occurs somewhere within the new batch, the latter will still appear to have drifted from the current example set [1].

Mining with the TSAR approach produces a knowledge structure in which individual rules may be tagged with, i.e. have as part of their antecedent, a time stamp 'current', a time stamp 'new' or no stamp at all. Any rule having time stamp value of 'current' must be out of date since here 'current' refers to the situation before the present round of mining. Such rules are called *invalid* while the remaining rules are *valid* [1].

The final stage in a CD3 cycle involves updating the database of current examples by first purging it of those examples that match invalid rules and then adding the new batch to it after re-labelling its examples as current. This process is explained at greater length in [1].

Valid rules are used for the on-line classification task until they are further updated after the next round of mining.

We demonstrated in [1] that CD3 could detect revolutionary and evolutionary patterns of drift against a substantial background of noise and recover swiftly to produce effective classification. This was in marked contrast to the use of ID3 where time stamps were not used.

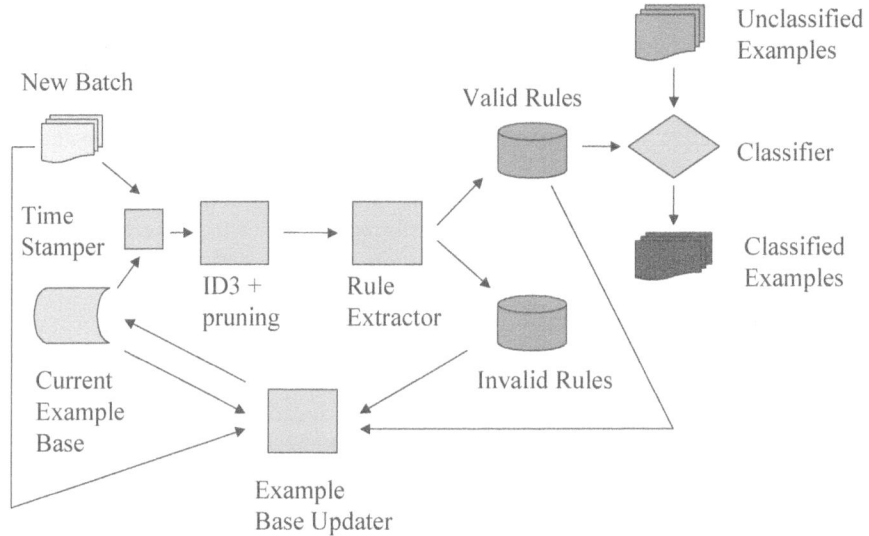

Fig. 1. The CD3 algorithm

3   Refining the Time Stamps

CD3 uses a very simple form of time stamping relying on just two values - 'current' and 'new'. This is sufficient to allow it to separate valid and invalid rules and to maintain good classification rates using the former. One possible disadvantage, however, is that as mining proceeds, the examples purged in previous rounds are lost to the system. This is the case even though such purges may be false, i.e. may have occurred as a result of errors in the induction process. It was shown in [1] that false purging even under realistic noise levels could be kept to a minimum. Nevertheless, it is worth considering how the mining process could be allowed to review and maybe revoke earlier decisions.

3.1   Time Stamping with Batch Identifiers: The CD4 Algorithm

To achieve this requires more refined time stamping. As a first possibility, each batch is assigned and retains indefinitely its own batch identifier. At each round of mining, all the data from the previous batches is used together with the new batch. There is no purging process. Instead the base learning algorithm is able to distinguish valid and invalid rules by appropriate instantiation of the time stamp attribute possibly revising decisions made in a previous induction.

To make matters concrete, suppose we use ID3 with post-pruning as the base learner, as in CD3. The time stamp attribute, say ts, which we can assume to be numeric, e.g. just batch number, can be designated as continuous and thus subject to binary splits of form $ts \leq T$ versus $ts > T$. Through selection of appropriate values of T, the algorithm is able to identify possible drift point(s). Typically an induced tree may have many such ts split nodes, some just reflecting noise.

Valid and invalid rules can be extracted from the tree as for CD3. Any tree path identifying $ts \leq T'$ where T' is less than the most recent value of the ts (that of the new batch) maps to an invalid rule. All other paths map to valid rules.

As before, valid rules would be used for on-line classification. If a collection of currently valid examples is required, say for boosting to improve classification rate, this can be extracted in the same manner as that used for purging. Unlike with CD3, however, this process is not an integral part of the induction mechanism.

We refer to the procedure just described as the CD4 algorithm. As with CD3, the decision tree inducer and the pruning mechanism are parameters of the algorithm.

3.2   Removing Batch Effects with Example-Based Time Stamping: The CD5 Algorithm

Although the unit of incrementation of mining is the batch, we can remove the effects of batches on mining by using continuous time stamping in which each training example is has its own unique time stamp. This would be a numeric attribute. The base learning algorithm is now free to form binary splits as for CD4 but without regard to batch. Thus it can place a split point within a batch (either the new or any of the previous batches) and review these decisions at each new round of mining.

Again the procedures for extraction of valid and invalid rules and maintenance of a database of currently valid examples are as described above. As with CD4, purging is not an integral part of the incremental mining process.

We refer to this example-based time stamping mining procedure as CD5.

3.3   Comparing CD3 with CD4 and CD5

We motivated the discussion on refined time stamping by arguing that it allowed us to avoid the purging process in CD3. The latter involves irredeemable loss of examples and so may appear undesirable. In contrast, though, CD4 and CD5 both retain all examples indefinitely. Whilst this allows for re-assessment of when (and if) drift occurred, it can create new problems.

With many time stamp values, i.e. potential split points, the algorithms are given a greater opportunity to infer drift points where they may not exist. This may lead to an increase in the number of spurious lower level splits that resist the pruning algorithm.

CD4 and CD5 decision trees also present a greater problem of interpretation. As mining proceeds, the trees will become larger and, perhaps unwieldy. There may be many different ts values identified in split points. In addition to split points involving, for CD4, the most recent batch and, for CD5 a batch identifier within the new batch, there may be 'old' split points identifying drift in one or more previous batches. If these have been identified for the first time, then their validity may be in doubt.

Anyway, depending on the domain, such retrospective identification may be of little use.

The issue here, then, is one of acuity of the drift detection mechanism and about which of the three algorithms delivers the best detection. This may depend on the domain and also on factors such as batch size.

An additional concern is that at each round of mining the data set being used will tend to be larger for CD4 and CD5 than for CD3. In domains with a very frequent need for on-line classification, speed of learning may be an important factor.

In CD3, the main indicators of drift are the highest position achieved by the time stamp attribute in the ID3 tree and the percentage of the current example base that is purged. At a drift point, the high position rises markedly in the tree and then falls quickly (unless drift is evolutionary); the percentage purged rises from its background false purging level before falling again [1]. In contrast, with CD4 and CD5 we can expect the high time stamp split to maintain its position over future batches. This is because a true drift point, once identified, should always be present as a discriminator of data before and after it.

4   Experimental Comparison of CD3, CD4, and CD5

To investigate how CD4 and CD5 respond to drift of various types and to compare their performance to CD3, we performed a number of experiments in which each of the three algorithms were run on the same data sets. Each experiment was replicated 10 times. We used data generated from the artificial domain described in detail in [1]. This has nine description attributes for examples, three of which are pure noise. There are three classes. There are two rule sets one applying before drift (Univ1) and one applying after (revolutionary) drift (Univ2). Evolutionary drift is simulated by a gradual move from Univ1 to Univ2 using mixing probabilities. We used the highest of the three noise levels in [1], referred to there as moderate noise.

Because our ID3 and post pruner used in [1] for the CD3 implementation does not cater for continuous attributes, we were obliged to find another base learner for CD4 and CD5. We chose C5, the successor to C4.5 [10], and used the default settings. This will present a difficulty when we compare the algorithms because any differences in performance might be due, at least in part, to that of the base learners. Nevertheless our purpose here is to see whether the refined time stamps can offer comparable performance to that CD3. A more thorough comparison using identical base learners will be undertaken in the future.

In addition, to illustrate the consequences of ignoring time stamps we give results for C5 applied to the totally of data available at the current time point and to C5 applied to just the new batch referred to as C5nb. The latter shows what would be achieved by forgetting all old data on the receipt of a new batch. This can be regarded as a simple-minded strategy for dealing with drift.

Results are presented in figures 2 (a), 3, 4 and 5 as learning curves averaged, over the ten trials, showing the actual classification rate (ACR) against the time point, n, which for convenience, is represented as the total number of examples fed to the system so far. The ACR [1] is the true classification rate of the valid rules applied to the rule set in force as calculated from the known probability distributions for the domain.

4.1   CD3 vs. CD4: Revolutionary Drift between Batches

Initially we compared CD3 and CD4 on revolutionary drift. Data was presented to the algorithm in ten batches of 100, with drift occurring immediately after the fifth batch. Thus in total, 1000 examples were presented to the learner in each trial. For CD4, the first 100 examples had batch-id 100, the next 100 had batch-id 200 and so on.

Figure 2 (a) shows the ACR for CD3, CD4 and C5nb before, during and after drift. Both CD3 and CD4 are able to detect the drift and continue to provide an acceptable ACR. CD4 has a slight advantage over CD3 before drift, however this does not persist during or after drift. This may be explained by the different pruners of the two base algorithms with CD3 not removing the ts attribute as effectively in smaller sample sizes. The performance of both CD algorithms does fall below that of C5nb at the drift point. This is quickly rectified with the next batch update. CD3 and CD4 continue to rise above C5nb thereafter. Note that the ACR for C5nb rises after the drift point solely because the drifted rules are simpler and therefore easier to mine. Beyond the drift point, the enhanced time stamping seems to afford very little advantage to CD4. Put another way, our anxiety about the detrimental affect of false purging in CD3 seems to be unfounded.

Figure 2 (b) shows the average highest position of the ts attribute within the CD3 and CD4 trees measured as depth from the root (the root position is zero with lower nodes being represented by higher values). That drift occurs is very clearly indicated by the behaviour of the ts attribute. For CD3 we see the characteristic trough at the drift point as the ts attribute temporarily reaches the higher parts of the tree only to fall again [1]. For CD4, the effect is different. There is a similar rise in ts at the drift point but thereafter, the value hovers at this high position as anticipated in the discussion above. Both algorithms are, however, in slightly different ways, quickly registering that drift has occurred.

4.2   CD4 vs. CD5: Revolutionary Drift between Batches

We now compared CD4 and CD5. We used a longer time span – to 2000 examples again on batches of 100. Data is presented to the algorithms in the same batch wise manner but for CD4 the ts attribute is a batch-id with 20 discrete values in multiples of 100: 100...2000, and in the other it is an example-id with 2000 continuous values 1...2000. C5 and C5nb were also included for comparison.

The results are shown in figure 3. All incremental algorithms have a similar ACR before drift. C5 performs best because there is no drift. CD5 is slightly lower than CD4. This may be because it has too much licence to infer ts split points as suggested above – especially dangerous at small sample sizes. But, when drift occurs, the two CD algorithms clearly show they have the advantage in detecting drift and in recovering. Subsequently they quickly rise above the C5nb level of performance. CD4 and CD5 recover initially at the same rate. CD5, however, takes a marginal lead a little further out the learning curve. It is interesting, and not clear why, CD5 should have an advantage over CD4 far from a single drift point at 500.

When C5 encounters drift, without the aid of the ts attribute, it rapidly deteriorates and its recovery is very slow.

4.3   CD4 vs. CD5: Revolutionary Drift within a Batch

In the scenarios so far we have looked at revolutionary drift occurring between batches. However, depending on the collection of the data, drift may occur as revolutionary within a batch.

The next experiment was carried out using the CD algorithms with the batch sizes doubled giving 10 batches of size 200. This allows drift to occur within the 3rd batch, still after time point n=500.

We see from figure 4 that, at the drift point, the fall in ACR for CD5 is much less that that for CD4. Here the ability of CD5 to determine split anywhere within a batch is paying off. Given that in many applications it is likely that drift occurs within a batch, or, rather, it cannot be asserted that it will not, this is an important feature of the algorithm. At other points on the learning curve the performances are as noted above for end of batch drift: a slight advantage to CD4 at smaller sample sizes and the reverse at larger sizes.

4.4   Evolutionary Drift over a Number of Batches

Evolutionary drift over a number batches is arguably the difficult situation in which to mine. The rate of the drift will determine how well the learner will detect it. We simulated evolutionary drift, as in [1], by moving from a mixture ratio of *Univ1:Univ2* at 100:0 to 90:10 to 80:20 and so on until we will have totally drifted to Univ2 (a mixture ratio 0:100). We refer to this as evolutionary drift at a rate of 10% per batch. Drift begins after time point n=500 and occurs at every time point thereafter completing at time point n=1500. All three CD algorithms were used.

Figure 5 clearly highlights the difficulty that evolutionary drift presents especially when the rate of drift is slow, extending over a long time period, and against a noisy background. When drift begins, all algorithms experience a considerable drop in ACR, larger than seen in previous drift scenarios, and recovery through the drift period is slow. The three CD algorithms have a similar recovery rate with CD4 and CD5 showing a only marginal benefit over CD3.

As shown in [1], learning, using only the new batch, performs very badly in situations of evolutionary drift taking a very long time to recover.

5   Conclusions and Further Work

We have extended the use of time stamps in TSAR algorithms beyond the two-valued attribute used in CD3. Both the extension to individual batch time stamps and further to individual example time stamps in algorithms CD4 and CD5 respectively appear to produce results comparable to, and possibly, slightly superior to those obtained from CD3. Further experiments using a common base decision tree learner will be necessary to confirm this. All three CD algorithms are superior to C5 and C5nb. C5, used without the benefit of time stamps and thus the capacity to distinguish valid and invalid rules, suffers substantial falls in classification rate when drift occurs. The simple strategy of dealing with drift by using just the new batch data is effective only immediately after drift. Elsewhere it prevents growth in ACR through accumulation of data. It also, of course, denies us the opportunity to detect drift should it occur.

If the benefit from the enhanced time stamps turns out to be marginal then the choice of which algorithm to deploy may depend on the characteristics of the domain used and the nature of the on-line performance task. CD4 and CD5 will produce larger trees than CD3 and take slightly longer to learn.

In the TSAR approach, we have been concerned with the rapid identification of drift (itself an important practical goal of mining) and with maintaining the success rate of the deployed classifier. We have not yet considered how the approach may be further harnessed to model the change in knowledge over time. Clearly an induced tree that contains time stamp information is itself such a model. In this regard, both CD4 and CD5 offer a benefit over CD3 since they induce trees that record the total history of changes in the underlying rules and therefore provide a basis for further analysis.

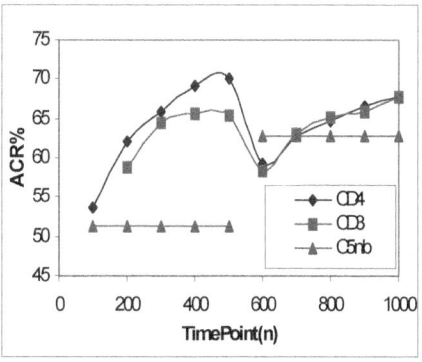
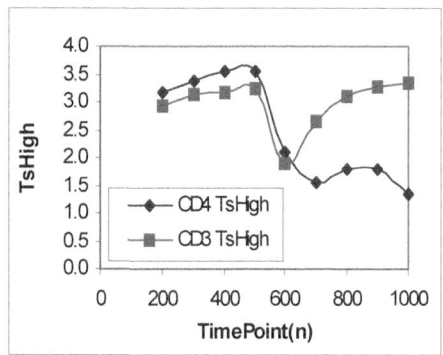

Fig. 2 (a). CD3 vs CD4

Fig. 2 (b). CD3 vs CD4 - *ts* High Position

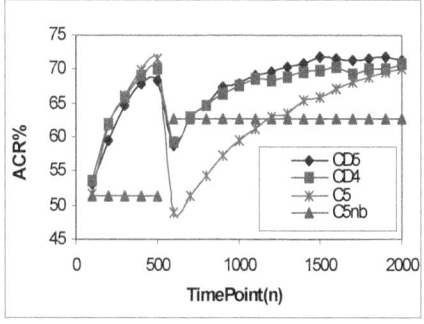
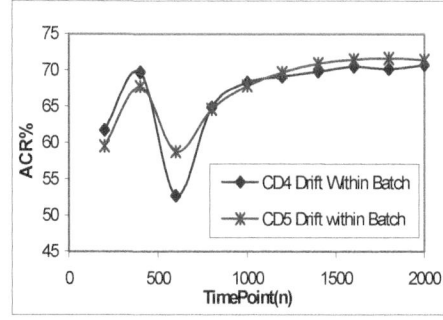

Fig. 3. CD4 vs CD5 – Drift between Batches

Fig. 4. CD4 vs CD5 – Drift within a Batch

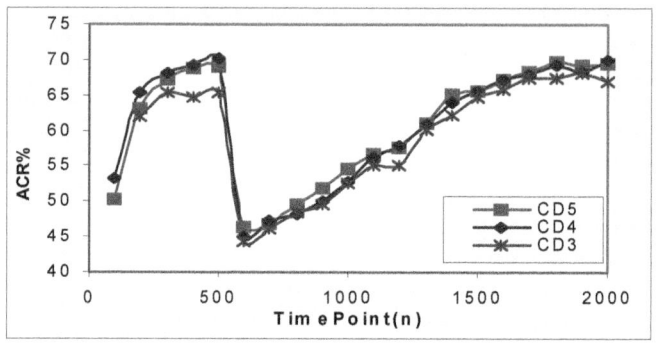

Fig. 5. CD3 vs CD4 vs CD5 for Evolutionary Drift @ 10%

References

1. Black, M., Hickey, R.J.: Maintaining the Performance of a Learned Classifier under Concept Drift. Intelligent Data Analysis 3 (1999) 453-474
2. Kelly, M.G., Hand, D.J., Adams, N.M.: The Impact of Changing Populations on Classifier Performance. In: Chaudhuri, S., Madigan, D. (eds.): Proceedings of the Fifth ACM SIGKDD International Conference on Knowledge Discovery and Data Mining. Association for Computing Machinery, New York (1999) 367-371
3. Klenner, M., Hahn, U.: (1994). Concept Versioning: A Methodology for Tracking Evolutionary Concept Drift in Dynamic Concept Systems. In: Proceedings of Eleventh European Conference on Artificial Intelligence, Wiley, Chichester, England, 473-477
4. Schlimmer, J.C., Granger, R.H.: Incremental Learning from Noisy Data. Machine Learning 1 (1986) 317-354
5. Hembold, D.P., Long, P.M.: Tracking Drifting Concepts by Minimising Disagreements. Machine Learning 14 (1994) 27-45
6. Widmer, G.: Tracking Changes through Meta-Learning. Machine Learning 27 (1997) 259-286
7. Widmer, G., Kubat, M.: Learning in the Presence of Concept Drift and Hidden Contexts. Machine Learning 23 (1996) 69-101
8. Chakrabarti, S., Sarawagi, S., Dom, B.: Mining Surprising Patterns using Temporal Description Length. In: Gupta, A., Shmueli, O., Widom, J. (eds.): Proceedings of the Twenty-Fourth International Conference on Very Large databases. Morgan Kaufmann, San Mateo, California (1998) 606-61
9. Chen, X., Petrounias, I.: Mining Temporal Features in Association Rules. In: Zytkow, J., Rauch, J. (eds,): Proceedings. of the Third European Conference on Principles and Practice of Knowledge Discovery in Databases. Lecture Notes in Artificial Intelligence, Vol. 1704. Springer-Verlag, Berlin Heidelberg New York (1999) 295-300
10. Quinlan, J.R.: C4.5: Programs for Machine Learning. Morgan Kaufmann, San Mateo, California (1993)

K-Harmonic Means
-A Spatial Clustering Algorithm with Boosting

Bin Zhang[1], Meichun Hsu, Umesh Dayal
Hewlett-Packard Research Laboratory
{bzhang, mhsu, udayal}@hpl.hp.com

Abstract. We propose a new center-based iterative clustering algorithm, K-Harmonic Means (KHM), which is essentially insensitive to the initialization of the centers, demonstrated through a set of experiments. The dependency of the K-Means performance on the initialization of the centers has been a major problem; a similar issue exists for an alternative algorithm, Expectation Maximization (EM). Many have tried to generate good initializations to solve the sensitivity problem. KHM addresses the intrinsic problem by replacing the *minimum* distance from a data point to the centers, used in K-means, by the Harmonic Averages of the distances from the data point to all centers. KHM significantly improves the quality of clustering results comparing with both K-Means and EM. The KHM algorithm has been implemented in both sequential and parallel languages and tested on hundreds of randomly generated datasets with different data distribution and clustering characteristics.

Keywords. Clustering, K-Means, K-Harmonic Means, Data Mining.

1. Introduction

Clustering is one of the principle workhorse techniques in the field of data mining [FPU96], statistical data analysis [KR90], data compression and vector quantization [GG92], and many others. K-Means (*KM*), first developed more than three decades ago [M67], and the Expectation Maximization (*EM*) with linear mixing of Gaussian distributions [DLR77] are two of the most popular clustering algorithms [BFR98a], [SI84], [MK97]. See [GG92] for more complete references for K-Means and [MK97][RW84] for EM.

K-Means stands out, among the many clustering algorithms developed in the last few decades, as one of the few most popular algorithms accepted by many application domains. However, K-Means does have a widely known problem – the local optimum it converges to is very sensitive to the initialization. Many people have proposed initialization algorithms. We have an example to show that a seemingly very good initialization may not lead K-Means to a better local optimal than a random initialization (See **Fig. 1** below). How to initialize K-Means is not well understood.

In the following example, K-Means is setup to find 100 clusters of the BIRCH data set (from UC Irvine) composed from 100 normally distributed local clusters, in a

---

[1] Primary Contact: bzhang@hpl.hp.com. This document is released as a technical report in Oct. 1999, available at http://www.hpl.hp.com/techreports/1999/HPL-1999-124.html.

J.F. Roddick and K. Hornsby (Eds.): TSDM 2000, LNAI 2007, pp. 31–45, 2001.
© Springer-Verlag Berlin Heidelberg 2001

10x10 grid, each has 1000 points. Two experiments were conducted – one with a random initialization and the other with an initialization generated by the Furthest Point algorithm [G85], which by itself is considered a clustering algorithm. Both initializations have exactly 100 centers (drawn as black "x" in **Fig. 1**). The second initialization seems a lot better than the first one, but the number of pairs of centers trapped by local densities differs by one in the local optima K-Means converged to under both initializations. The best convergence (global optimum) should have exactly one center in each local cluster of the data.

Instead of inventing or improving an initialization for K-Means, we look into the intrinsic problem that resulted in K-Means sensitivity to initialization – its winner-takes-all partitioning strategy, which makes the association between data points and the nearest center so strong that the membership of a data point is not changed until it is closer to a different center. This strong association drags the centers from moving out of a local density of data. We use the association provided by the *harmonic means* function, to replace the winner-takes-all strategy of K-Means. With this new strategy, the association of the data points with the centers is distributed (like EM, but EM has certain problems, pointed out in Section 6 and **Fig. 4**, that prevent it from reaching a good clustering) and the transition becomes continuous. We also show that KHM algorithm has a "build-in" boosting function, which boosts the data that are not close to any center by giving a smaller weight on the data that are already close to one of the centers.

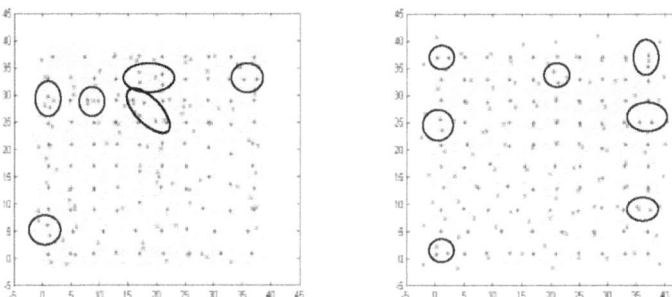

Fig. 1. Left: Random Initialization. Right: Initialization by the Furthest Point algorithm. Black "x's are initializations of the centers and red dots are converged results. There are 8 pairs of centers trapped by local densities of data in the converged results for both initializations.

We call this weighting function *a dynamic weighting function* because it is automatically adjusted in each iteration. With these changes, the new algorithm is essentially insensitive to initialization. We demonstrate that by starting KHM with very bad initializations and comparing its convergence with KM or EM that started with random initializations. Comparing all three algorithms (KM, KHM, and EM) under a unified view (see Section 7) gives a more detailed explanation of KHM's insensitivity to initialization.

1.1 The Performance Function of K-Harmonic Means

K-Means' performance function is

$$Perf_{KM}(\{x_i\}_{i=1}^N,\{m_l\}_{l=1}^K) = \sum_{l=1}^K \sum_{x \in S_l} \| x - m_l \|^2, \tag{1}$$

where $S_l \subset X = \{x_i\}_{i=1}^N$ is the subset of x's that are closer to m_l than any other centers in $M = \{m_l\}_{l=1}^K$. (Or $\{S_l | l=1,...,K\}$ is the Voronoi partition). The double summation in (1) can be considered as a single summation over all x (data points) and the squared distance under the summations can be expressed by *MIN()*. Therefore, the KM performance function can be rewritten as

$$Perf_{KM}(X,M) = \sum_{i=1}^N MIN\{\| x_i - m_l \|^2 | l = 1,...,K\}, \tag{2}$$

Replacing *MIN()* by *HA()*, we get the performance function of *KHM*:

$$Perf_{KHM}(X,M) = \sum_{i=1}^N HA\{\| x_i - m_l \|^2 | l = 1,...,K\} = \sum_{i=1}^N K \Big/ \sum_{l=1}^K \frac{1}{\| x_i - m_l \|^2}. \tag{3}$$

The quantity inside the outer summation is the harmonic average of K squared distances $\{\|x - m_l\|^2 \mid l = 1,...,K\}$. To give a reason for choosing this function, we briefly review the concept of *harmonic average* (also called harmonic mean) in the next section.

A unified view of the KM', KHM' and EM's performance functions is given later in Section 5 and 6, in which all are considered as ways of mixing bell-shape functions. Linear mixing of bell-shape functions is used in EM [MK97].

The rest of the paper is organized as follows: Harmonic Average *HA()*; comparing *HA()* with *MIN()*; KHM' algorithm; implementation issues; computational complexity per iteration; EM algorithm based on linear mixing (limited version); a unified view of all three performance functions – KM', KHM' and EM's; different ways of mixing bell shape functions; a unified view of all three algorithms; experimental results.

2. Harmonic Average Function and Comparison with *MIN()*

The harmonic average of K numbers $\{a_i \mid i = 1,...,K\}$ is

$$HA(\{a_i \mid i = 1,...,K\}) = K \Big/ \sum_{i=1}^K \frac{1}{a_i}. \tag{4}$$

Fig. 2 has a plot of *MIN(X,Y)* and *HA(X,Y)/2*, (X,Y) in [0,10]x[0,10]. The two horizontal axis are X and Y; the vertical axis is the value of the functions. The plot of *HA()* is very similar to that of *MIN()*. More detailed mathematical comparisons of the two functions are possible based on their boundary conditions and their derivatives. Due to the limited length of this paper, we omit it.

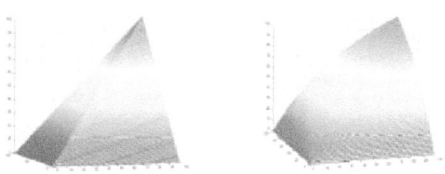

Fig. 2. The plots of *MIN()* on the left and *HA()/K* on the right.

3. The *KHM* Algorithm

To derive an optimization algorithm for minimizing the KHM performance function, we take partial derivatives of the KHM's performance function (3) with respect to the center positions m_k, $k=1,...,K$, and set them to zero,

$$\frac{\partial Perf_{KHM}(X,M)}{\partial m_k} = -K * \sum_{i=1}^{N} \frac{2*(x_i - m_k)}{d_{i,k}^4 (\sum_{l=1}^{K} \frac{1}{d_{i,l}^2})^2} = \vec{0} \qquad (5)$$

where $d_{i,l} = ||x_i - m_l||$ on the right of (5) are still functions of the centers. "Solving" m_k's from the last set of equations, we get a recursive formula:

$$m_k = \sum_{i=1}^{N} \frac{1}{d_{i,k}^4 (\sum_{l=1}^{K} \frac{1}{d_{i,l}^2})^2} x_i \Bigg/ \sum_{i=1}^{N} \frac{1}{d_{i,k}^4 (\sum_{l=1}^{K} \frac{1}{d_{i,l}^2})^2}. \qquad (6)$$

This is the formula for KHM' recursion. KHM, like KM and EM, is a also center-based, iterative algorithm that refines the clusters defined by the K centers. Starting with a set of initial positions of the centers, KHM calculates $d_{i,l} = ||x_i-m_l||$, and then the new positions of the centers from (6) or from the decomposed sequence below (implementation details are given later),

$$\alpha_i = 1/(\sum_{l=1}^{K} \frac{1}{d_{i,l}^2})^2, \quad q_{i,k} = \frac{\alpha_i}{d_{i,k}^4}, \quad q_k = \sum_{i=1}^{N} q_{i,k}, \quad p_{i,k} = \frac{q_{i,k}}{q_k}, \quad m_k = \sum_{i=1}^{N} p_{i,k} x_i. \qquad (7.1\text{-}7.5)$$

The recursion is continued until the performance value stabilizes.

4. The EM Algorithm Using Linear Mixing of Gaussian Distributions

We briefly review a version of the EM algorithm needed later for the comparison with KHM and KM. We limit ourselves to the EM algorithm with linear mixing of K identical spherical bell-shape (Gaussian distribution) functions, which matches the bell-shape functions used in KM and KHM.

Let

$$Perf_{EM}(X,M) = -\log\left\{\prod_{i=1}^{N}[\sum_{l=1}^{K} p_l * \frac{1}{\sqrt{(2\pi)^D}} EXP(-\|x-m\|^2)]\right\}. \tag{8}$$

a linear mixing of K identical spherical bell-shape functions. EM algorithm is a recursive algorithm with the following two steps:

E-Step:
$$p(m_l \mid x_i) = \frac{p(x_i \mid m_l) * p(m_l)}{\sum_{i=1}^{N} p(x_i \mid m_l) * p(m_l)}, \tag{9}$$

where $p(x|m)$ is the prior probability with Gaussian distribution, $p(m_l)$ is the mixing probability.

M-Step: $m_l = \sum_{i=1}^{N} p(m_l \mid x_i) * x_i \Big/ \sum_{i=1}^{N} p(m_l \mid x_i)$, and $p(m_l) = \frac{1}{N}\sum_{i=1}^{N} p(m_l \mid x_i)$, (10) and (11)

where N is the size of the whole data set. For more details, see [MK97] and the references there.

5. A Unified View of the Three Performance Functions

Without introducing any change, applying the identity mapping $-log(EXP(-(\ )))$ to the performance functions of KM and KHM, we get

$$Perf_{KM}(X,M) = -\log(\prod_{i=1}^{N} EXP(-MIN\{\|x-m\|^2 \mid m \in M\})); \tag{12}$$

$$Perf_{KHM}(X,M) = -\log(\prod_{i=1}^{N} EXP(-HA\{\|x-m\|^2 \mid m \in M\})). \tag{13}$$

Now they share the same form of the EM's performance function:

$$Perf_{EM}(X,M) = -\log(\prod_{i=1}^{N}[\sum_{l=1}^{K} p_l * \frac{1}{\sqrt{(2\pi)^D}} EXP(-\|x-m\|^2)]). \tag{14}$$

If we remove the negative sign in front of the "log", all three algorithms can be considered as maximizing the likelihood functions, except that $EXP(-MIN\{\|x-m\|^2 \mid m \in M\})$ in (12) and $EXP(-HA\{\|x-m\|^2 \mid m \in M\})$ in (13) are not normalized to be density functions. Therefore K-Means and K-Harmonic Means are not exactly EM-type of algorithms. The quantity inside the brackets "[]" in (14) is the linear mixing of the bell-shape functions – the $EXP()$'s. We can also look at the performance functions of KM and KHM as mixings of bell-shape functions.

6. MIN, Harmonic, and Linear Mixings of Bell-Shape Functions

The performance functions of KM and KHM can be viewed as a (log of) mixing of bell-shape functions.

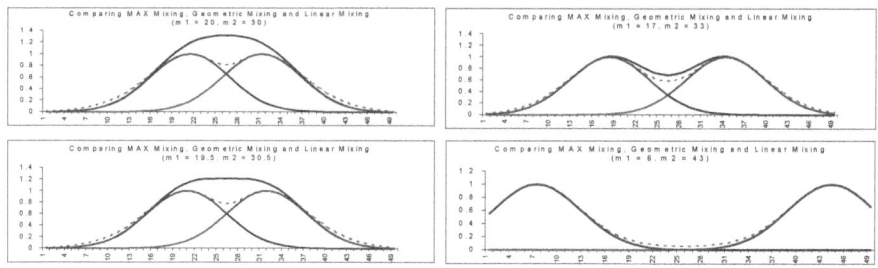

Fig. 4. Comparing Three Different Mixings of Two One-dimensional Bell-shape Functions. The function value between centers from low to high: *MIN Mixing in KM* (hard to see because it perfectly overlap with the individual bells), Harmonic Mixing in KHM, and Linear Mixing in EM. As the centers move from near to far, the difference among the three mixings decrease. In the plot at the lower-right corner, all three mixings are almost identical.

Define

MIN Mixing: $EXP(-MIN\{\|x-m_l\|^2 \mid l=1,...,K\})$ (15)

Min-Mixing can also be called Max-Mixing because *EXP(-x)* is monotone decreasing and

$$EXP(-MIN\{\|x-m_l\|^2 \mid l=1,...,K\}) = MAX\{EXP(-\|x-m\|^2) \mid l=1,...,K\}. \quad (16)$$

Harmonic Mixing: $EXP(-HA\{\|x-m_l\|^2 \mid l=1,...,K\})$ (17)

Linear Mixing: $\displaystyle\sum_{l=1}^{K} p_l * \frac{1}{\sqrt{(2\pi)^D}} EXP(-\|x-m_l\|^2)$ (18)

For finding clusters, linear mixing (the limited version with fixed covariance matrix of the bells) does not behave properly when centers get too close to each other. As two or more bell-shape functions merge into a single peak, the maximum of that single peak behaves like a ghost center and the individual centers loose their identity. This is clearly shown in our experimental results (See the experimental results from EM in Group 2 and Group 4).

7. A Unified View of Three Algorithms – KM, KHM, and EM

We compared three performance functions in the last section. In this section, we compare all three algorithms.   All three algorithms take the following form,

$$m_l = \frac{\displaystyle\sum_{i=1}^{N} p(m_l \mid x_i) * a(x_i) x_i}{\displaystyle\sum_{i=1}^{N} p(m_l \mid x_i) * a(x_i)}, \quad p(m_l \mid x_i) \geq 0 \text{ and } \sum_{l=1}^{K} p(m_l \mid x_i) = 1, \quad (19)$$

where $i=1, \ldots K$ and $a(x) \quad 0$, the weighting function of the data points, which decides how much of each data point x participates in the calculation of the new center locations. $p(m_l \mid x_i)$'s are the "membership" functions -- $p(m_l \mid x_i)$ portion of $a(x_i)*x_i$ is associated with m_l. For K-Means, all data points are equally weighted and the each data point belongs to the closest center 100% (winner-takes-all). Therefore, $p(m_l|x_i)=1$ if m_l is the closest center to x_i, otherwise 0. Ties are resolved arbitrarily. The weighting function $a(x) = 1$. For EM: all data points are equally weighted and the membership function is derived from Bayes' rule, where $p(x_i| m_l)$ is the lth bell-shape function and $p(m_l)$ the weight of the lth bell-shape function, (See (8) and (9) in Section 4.)[2]

$$p(m_l \mid x_i) = \frac{p(x_i \mid m_l)* p(m_l)}{\sum_{i=1}^{N} p(x_i \mid m_l)* p(m_l)}, \quad and \quad a(x) = 1.$$

For K-Harmonic Means, the iterative procedure (6) (in Section 3) can be written as

$$\vec{m}_k = \sum_{i=1}^{N} \frac{1}{d_{i,k}^{4} (\sum_{l=1}^{K} \frac{1}{d_{i,l}^{2}})^2} \vec{x}_i \Bigg/ \sum_{i=1}^{N} \frac{1}{d_{i,k}^{4} (\sum_{l=1}^{K} \frac{1}{d_{i,l}^{2}})^2} = \sum_{i=1}^{N} \frac{\frac{1}{d_{i,k}^{4}}}{\sum_{l=1}^{K} \frac{1}{d_{i,l}^{4}}} * \frac{\sum_{l=1}^{K} \frac{1}{d_{i,l}^{4}}}{(\sum_{l=1}^{K} \frac{1}{d_{i,l}^{2}})^2} * \vec{x}_i \Bigg/ \sum_{i=1}^{N} \frac{\frac{1}{d_{i,k}^{4}}}{\sum_{l=1}^{K} \frac{1}{d_{i,l}^{4}}} * \frac{\sum_{k=1}^{K} \frac{1}{d_{i,l}^{4}}}{(\sum_{l=1}^{K} \frac{1}{d_{i,l}^{2}})^2}$$

where $d_{i,k}=||x_i-m_k||$. For KHM, we have

$$p(m_k \mid x_i) = \frac{\frac{1}{d_{i,k}^{4}}}{\sum_{l=1}^{K} \frac{1}{d_{i,l}^{4}}} \quad and \quad a_M(x) = \frac{\sum_{l=1}^{K} \frac{1}{||x - m_l||^4}}{[\sum_{l=1}^{K} \frac{1}{||x - m_l||^2}]^2}.$$

8. Dynamic Weighting of Data -- Boosting for Unsupervised Learning

The weighting function $a(x)$, when designed properly, significantly reduces the sensitivity of the convergence quality to the initialization of the centers. This is done by defining a generalized KHM as

$$\vec{m}_k = \sum_{i=1}^{N} \frac{1}{d_{i,k}^{s} (\sum_{l=1}^{K} \frac{1}{d_{i,l}^{2}})^2} \vec{x}_i \Bigg/ \sum_{i=1}^{N} \frac{1}{d_{i,k}^{s} (\sum_{l=1}^{K} \frac{1}{d_{i,l}^{2}})^2}.$$

---

[2] The decomposition is not unique. But for the discussion here, a decomposition is sufficient.

For the generalized KHM, we have

$$p(m_k \mid x_i) = \frac{\dfrac{1}{d_{i,k}^{\,s}}}{\displaystyle\sum_{l=1}^{K}\dfrac{1}{d_{i,l}^{\,s}}} \quad and \quad a_M(x) = \frac{\displaystyle\sum_{l=1}^{K}\dfrac{1}{\|x-m_l\|^s}}{[\displaystyle\sum_{l=1}^{K}\dfrac{1}{\|x-m_l\|^2}]^2} = \|x-m_{\min}\|^{4-s}\; \frac{\displaystyle\sum_{l=1}^{K}\left(\dfrac{\|x-m_{\min}\|}{\|x-m_l\|}\right)^s}{\left[\displaystyle\sum_{l=1}^{K}\left(\dfrac{\|x-m_{\min}\|}{\|x-m_l\|}\right)^2\right]^2}.$$

(Note: $a()$ does depend on M as a whole.) The first factor $\|x-m_{\min_x}\|^{4-s}$ in $a(x)$ forces the weight to zero as x is approached by a center. The more centers are near a data point the smaller the weight for that data point. This has the effect of flatten out a local density that trapped more than one centers. This is the most important difference between K-Harmonic Means and K-Means (or EM). Based on the fact that the weight of each data point in the calculation of the center locations in the next iteration depends on the current location of the centers, we call this dynamic weighting of the data points. Weighting the data points that are close to the centers have the same effect as boosting the data points that are not. Therefore, the dynamic weighting of data is boosting.

Looking at the experimental results on the birch data set with 100 centers by K-Means or by EM, there are quite a few local densities that trapped two centers. Lowering the weight of the data points near those centers (in the KHM) will help the centers to escape from the trap. **Fig. 5** gives a plot of the weighting function, $a(x)$, for two centers in one dimensional space with $s=3$ (this value is used in all the experiments). The two centers are $m_1=4$ and $m_2=16$ for the left figure and $m_1=7$ and $m_2=11$ for the right.

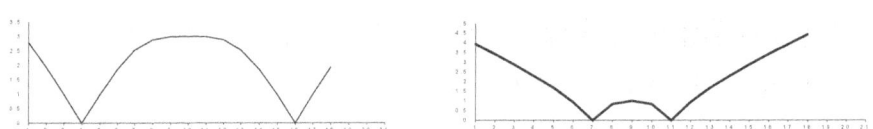

Fig. 5. A Plot of $a(x)$ for K-Harmonic Means with two centers in one-dimensional space. Left: when two centers are far. Right: when two centers are close.

9. Computational Costs of KHM in Each Iteration

In each iteration, calculating all the pair-wise distances from N data points to K centers (of D dimensional vectors) costs $O(N*K*D)$. KM and EM (linear mixing) share the same cost on this part. After getting the coefficients $p_{i,k}$, calculating the linear combinations $m_k = \sphericalangle\, p_{i,k}*x_i$ costs another $O(N*K*D)$. EM costs the same on this part. KM costs less ($O(N*D)$) on this due to the partitioning but an additional $O(N*K)$ comparisons and assignments (marking) are used to do the partitioning. After calculating the distances, all quantities used in the algorithm no longer depend on the dimension and all other costs are $O(N*K)$. The leading asymptotic term for all three algorithms are the same $O(N*K*D)$.

The asymptotic computational complexity *per iteration* for KM, KHM and EM (linear mixing model) are all $O(N*K*D)$. For all three algorithms, since the costs are dominated (especially for high dimensional data) by the distance calculations $||x_i-m_k||$, and there are exactly the same number of distances to be calculated, the coefficients of the cost term $N*K*D$ of all three algorithms are very close. It is the convergence rate and the convergence quality that[3] differentiate them in real world applications.

Space complexity of KHM is NxD for data points, KxD for the K centers and $KxD+2*K$ for temporary storage. The temporary storage requirement tend to be lower than KM because the later needs a $O(N)$ temporary storage to keep the membership information and $N>>K$ in real problems.

10. Implementation of KHM

The calculation of $q_{i,k}$'s (combination of (7.1) and (7.2)) are done as follows:

$$q_{i,k} = \frac{d_{i,\min}^{4}}{d_{i,k}^{s}[1+\sum_{l\neq \min}(\frac{d_{i,\min}}{d_{i,l}})^2]^2} = d_{i,\min}^{4-s}(\frac{d_{i,\min}}{d_{i,k}})^s \bigg/ [1+\sum_{l\neq \min}(\frac{d_{i,\min}}{d_{i,l}})^2]^2 \qquad (20)$$

where $d_{i,\min} = MIN(d_{i,l} \,|\, l = 1,...,K)$. All the ratios $(d_{i,\min}/d_{i,l})$ are in [0,1]. We have implemented KHM in several different languages -- C, Matlab and the parallel programming language ZPL (KHM has been run on multiple processors in parallel. See [ZHF]). We have tested KHM on hundreds of randomly generated datasets without encountering any numerical difficulties.

11. Experimental Results

Due to the limited length of this paper, we are not able to show all experimental results[4]. We focus on the experiments related to the issue of sensitivity to initialization in this section. Taking *s=3* in the KHM algorithm, we demonstrate that KHM converges to a much better local optimal from a much worse initialization comparing with K-Means and EM[5]. Two data sets are used: the BIRCH data set is from UC Irvine [B99]; and another data set, Hier, is generated by a Hierarchical Data Generator (HDG). The detailed information on these data sets is given in Table 1.

The HDG maps each point in a dataset into a cluster of points in either uniform or normal distribution with a variance chosen by the user. When this mapping is repeatedly applied, the clustering of the final resulted data set shows a hierarchical structure.

---

[3] Due to the partitioning nature, faster algorithms/implementations have been designed for KM using trees to do spatial partition of either the centers or the data [GG92],[PM99].

[4] All experimental results will be published on our website.

[5] This performance has been observed consistently on hundreds of randomly generated datasets. The detailed behavior of all three algorithms were observed through animation of the complete convergence paths.

Table 1. The Datasets.

Name	Size	Structure
BIRCH	100,000	100 clusters in a 10x10 grid with 1000 points in each cluster in normal distribution. The radius of each cluster is sqrt(2) and the neighboring clusters are 4*sqrt(2) apart.
Hier	20,000	Started with 4 vertices of a square, each one is mapped into 100 points. Each of the 400 points is mapped into 50 points. Both mapping used uniform distribution

We choose 2-dimensional data in this presentation for the power of visualization. We also compared KHM with KM and EM on a large number of three-dimensional datasets (with visualization of the convergence paths). KHM showed better convergence (less sensitive to initialization) consistently through out all experiments, especially when the initialization of the centers is not very good.

Twelve experiments conducted are organized in four groups:

Group 1 shows the convergence quality of the three algorithms on the BIRCH data set when the "correct" number of centers is used. All three algorithms started from the same random initialization. From the plots (given at the end of this paper), eight pairs of centers are trapped under KM (two centers found in one cluster) after its convergence. Only one pair of centers is trapped under KHM. Four pairs are under EM. The results are comparable under many different random initializations used (only one set is presented here).

Group 2 shows that, when the initialization is very bad, both KM and EM converge very slowly (if converge to anything meaningful at all). All three algorithms started from the same bad initialization. KM moves the centers out "layer-by-layer". KHM converges very fast and reached a configuration that is close to the global optimum in about 40 iterations. EM does not work well when the centers are close to each other (the reason was given earlier – the individual bell shape functions loose their identity and are merged into a big bell shape function with a ghost center). Most of the 2000 centers merged with others (approximating multiple clusters by a single bell-shape function)[6].

Group 3 shows that KHM can start from a really bad initialization, in which 400 centers are linearly spread out on a small line segment outside the region occupied by the data. KHM converged nicely. KM and EM do not work under this bad initialization. A random initialization is used for them instead.

Group 4 repeat the same experiment as Group 1 but with a bad initialization instead of the random initialization. In the Bad Initialization, 100 centers are linearly spread out on a small line segment in the center of all data (See the first plot in Group 4). KHM converged nicely in 90 iterations with only two pairs trapped. KM and EM do not converge well even after 300 iterations. This experiment shows that the proven asymptotic convergence property (rate) of KM and EM does not tell us exactly

---

[6] We did not get better results by allowing the "diameter" (covariance matrix) of the bell-shape functions to change.

how the algorithm will behave when the initialization is far from any local optimal. This confirms the known fact that the (asymptotic) convergence rate of an algorithm ≠ the convergence speed of the algorithm in practice.

Table 2. The Setup of the Experiments.

grp #	Alg.	Data Set	#of Centers	Snapshots Taken at	Local Opt.?	Init.
1	*KM*	BIRCH	*100*	*50*	Yes	Random
	KHM			*50,100,120*	No*	
	EM			*50,100*	No*	
2	*KM*	BIRCH	*2000*	*20,60,100,200*	No	Bad Init.
	KHM			*5,10,40,200*	No*	
	EM			*20,60,100,200*	No	
3	*KM*	Hier	*400*	*45*	Yes	Random
	KHM			*15,25,50*	No*	Bad Init.
	EM			*100*	No*	Random
4	*KM*	BIRCH	*100*	*10,25,100,300*	No	Bad Init.
	KHM			*10,25,40,90*	No*	
	EM			*10,25,100,300*	No	

No* -- See the last column.

In all the figures at the end of this paper, we plot the dataset in yellow as the background[7] and overlay the centers on top in red. We animated the convergence path in all experiments for all three algorithms. Unfortunately, we can provide only a few snapshots of each animation sequence on paper. The last snapshot listed in Table 2 is taken when there is no more detectable movement of the centers. We also plot the initializations of the centers for all experiments. The title bar of each figure is encoded as: *Group#: Algorithm, Data Set, Number of Centers, Number of Iterations Done, Initialization.*

12. Future Work

One of the reviewers of this paper pointed out the existing work [BEF84] on Fuzzy-c-Means. The K-Harmonic Means performance function is simpler than Fuzzy-c-Means. The algorithm is also simpler. Due to limited time and space, detailed comparisons will be done in the future.

---

[7] The BIRCH data set is reduced by random sampling to *20,000* data points before plotting for better visibility of the clusters.

References

[A73] Anderberg, M. R. 1973. Cluster analysis for applications. Academic Press, New York. 35p.

[B99] Bay, S. D. (1999). The UCI KDD Archive [http://kdd.ics.uci.edu]. Irvine, CA: University of California, Department of Information and Computer Science.

[BEF84] Bezdek, Ehrlich, & Full, "FCM: THE FUZZY c-MEANS CLUSTERING ALGORITHM", Computers & Geosciences, v.10, pp.191-203 , 1984

[BFR98] Bradley, P., Fayyad, U. M., and Reina, C.A., "Scaling EM Clustering to Large Databases," MS Technical Report, 1998.

[BF98] Bradley, P., Fayyad, U. M., C.A., "Refining Initial Points for KM Clustering", MS Technical Report MSR-TR-98-36, May 1998.

[BFR98a]Bradley, P.,Fayyad, U.M., and Reina, C.A., "Scaling Clustering to Large Databases", KDD98, 1998.

[DH72] Duda, R., Hart, P., "Pattern Classification and Scene Analysis", John Wiley & Sons, 1972.

[DLR77]Dempster, A. P., Laird, N.M., and Rubin, D.B., "Miximum Likelyhood from Incomplete Data via the EM Algorithm", Journal of the Royal Statistical Society, Series B, 39(1):1-38, 1977.

[FPU96]Fayyad, U. M., Piatetsky-Shapiro, G. Smyth, P. and Uthurusamy, R., "Advances in Knowledge Discovery and Data Mining", AAAI Press 1996

[GG92] Gersho & Gray, "Vector Quantization and Signal Compression", KAP, 1992

[GMW85] Gill, P.E., Murray, W. and Wright, H.M., "Practical Optimization", Academic Press, 1981.

[G85] Gonzales, T.F., "Clustering to Minimize the Maximum Intercluster Distance", Theo. Comp. Sci. 38, p293-306, 1985.

[KR90] Kaufman, L. and Rousseeuw, P. J., "Finding Groups in Data : An Introduction to Cluster Analysis", John Wiley & Sons, 1990.

[M67] MacQueen, J. 1967. Some methods for classification and analysis of multivariate observations. Pp. 281-297 in: L. M. Le Cam & J. Neyman [eds.] Proceedings of the fifth Berkeley symposium on mathematical statistics and probability, Vol. 1. University of California Press, Berkeley. xvii + 666 p.

[MA] McKenzie, P. and Alder, M., "Initializing the EM Algorithm for Use in Gaussian Mixture Modeling", The Univ. of Western Australia, Center for Information Processing Systems, Manuscript.

[MK97]McLachlan, G. J. and Krishnan, T., "The EM Algorithm and Extensions.", John Wiley & Sons, Inc., 1997

[PM99] Pelleg, D. and Moore, A, "Accelerating Exact K-Means Algorithms with Geometric Reasoning", KDD-99, Proc. of the Fifth ACM SIGKDD Intern. Conf. On Knowledge Discovery and Data Mining, page 277-281.

[RW84]Rendner, R.A. and Walker, H.F., "Mixture Densities, Maximum Likelihood and The EM Algorithm", SIAM Review, vol. 26 #2, 1984.

[SI84] Sclim, S.Z. and Ismail, M.A., "K-Means-Type Algorithms: A Generalized Convergence Theorem and Characterization of Local Optimality", IEEE Trans. On PAMI-6, #1, 1984.

[ZHF] Zhang, B., Hsu, M., Forman, G., "Accurate Recasting of Parameter Estimation Algorithms using Sufficient Statistics for Efficient Parallel Speed-up", To appear in PKDD 2000, September, Lyon, France.

Group 1: #1. Random Initialization used for all three experiments in Group 1.   #2. KM converged to a local optimal in 50 iterations and stopped by itself. 7 pairs trapped by local densities. #3. EM after 50 iterations. #4 - EM after 100 iterations. 5 pairs trapped.

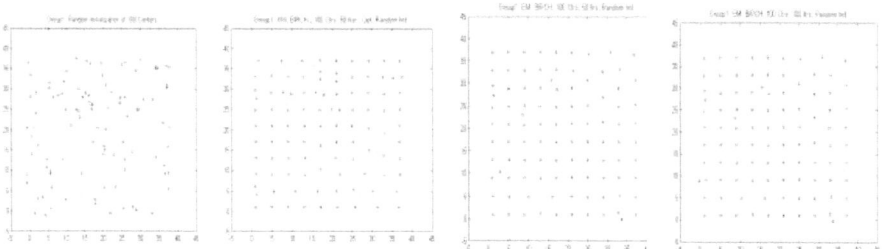

Group 1:#1. KHM after 50 iterations. #2. KHM after 100 iterations. #3. KHM after 120 iterations. Only one pairs of centers trapped by local densities.

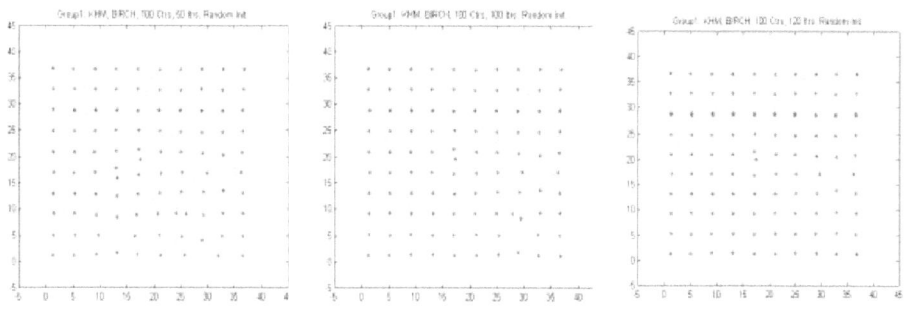

Group 2: The random initialization used for all experiments in Group 2. 2000 centers uniformly distributed around the center of the whole data set.

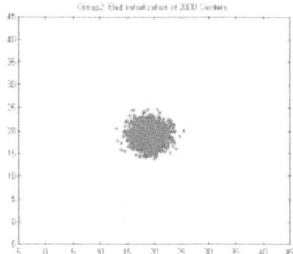

Group 2: #1. KM after 20 iterations. #2. KM after 60 iterations. #3. KM after 100 iterations. #4. KM after 300 iterations.

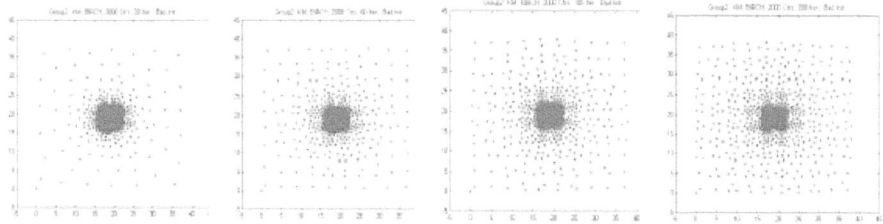

Group 2: #1. KHM after 5 iterations.   #2.   KHM after 10 iterations.   #3. KHM after 40 iterations.   #4. KHM after 200 iterations.   It takes only 40 iterations for the centers to be roughly distributed according to the data.

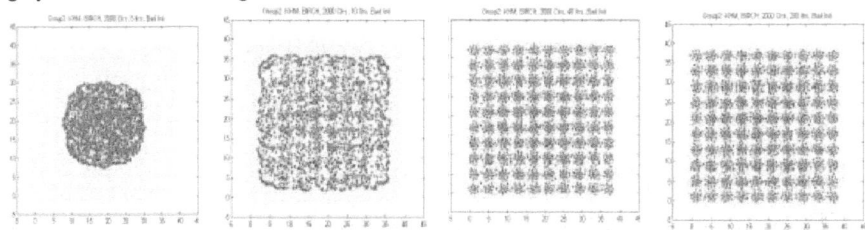

Group 2: #1 - EM after 20 iterations.   #2 - EM after 60 iterations.   #3 - EM after 100 iterations. #4 - EM after 300 iterations. This experiment clearly shows that the centers collapse into one another.   The individual bell-shape functions merge into a single big bell-shape function that tries to approximate the whole data dataset as a single normal distribution. Whether this is correct from the density approximation point of view can be further investigated.   But for the purpose of finding clusters, this result is not desirable.   This is explained in Section 6, **Fig.** 4. Different bell-shape functions merge and lose their individual identity.

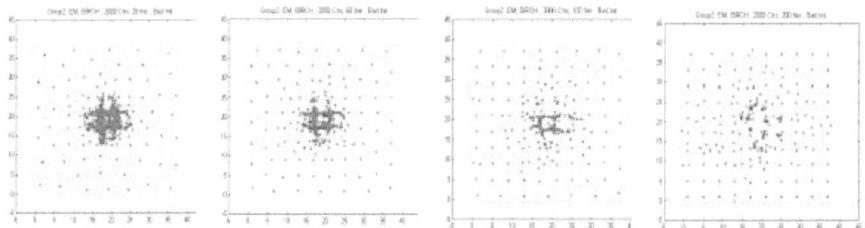

Group 3: #1 - Random initialization of 400 centers used for both KM and EM.   #2 - KM reached a local optimal after 45 iterations and stopped. #3 - EM after 100 iterations.

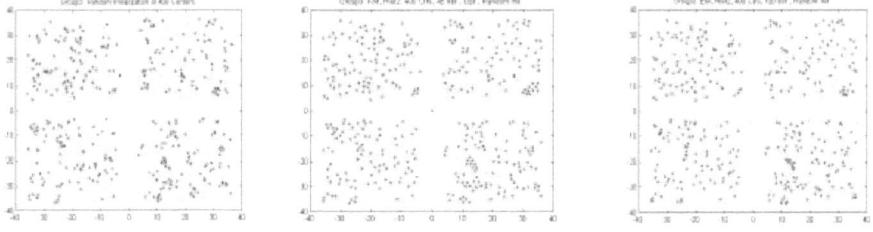

Group 3: #1. The bad initialization used for KHM in this experiment. #2. KHM after 15 iterations.   #3. KHM after 25 iterations.   #4. KHM after 50 iterations. The centers in the initialization fall outside the data region. Even with such a bad initialization, under which KM and EM will not work at all, KHM converged to a local optimal better than KM and EM give under the random initialization.

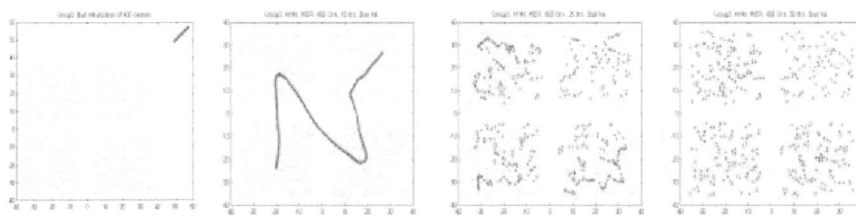

Group 4: The bad initialization used for all three experiments in this group. All 100 centers line up on a small segment near the center of the whole data set.

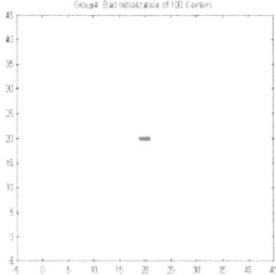

Group 4: #1. KM after 10 iterations. #2. KM after 25 iterations. #3. KM after 100 iterations. #4. KM after 300 iterations. Even after 300 iterations, KM does not converge well.

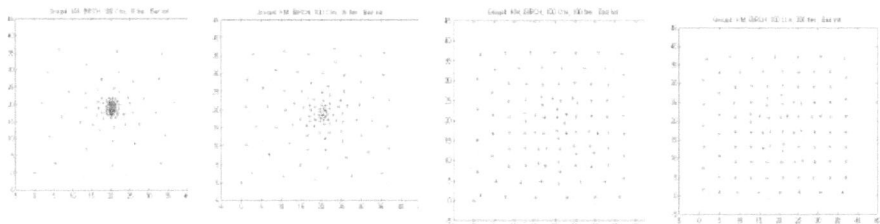

Group 4: #1. KHM after 10 iterations. #2. KHM after 25 iterations. #3. KHM after 40 iterations. #4. KHM after 90 iterations. It tooks only 90 iteration for KHM to stabilize. Only two pairs of centers are trapped.

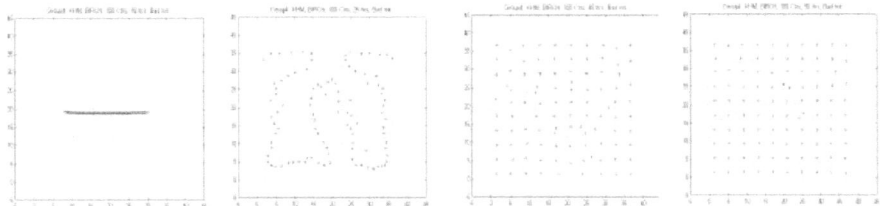

Group 4: #1. EM after 10 iterations. #2. EM after 25 iterations. #3. EM after 100 iterations. #4. EM after 300 iterations. EM does not converge very well after 300 iterations.

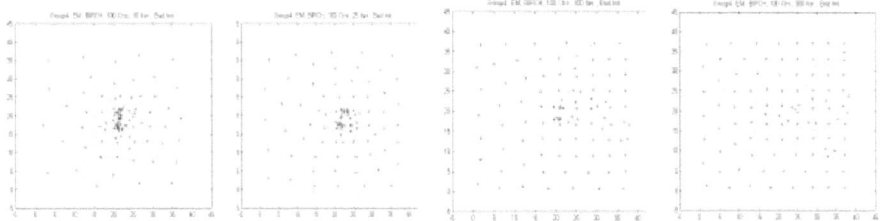

Identifying Temporal Patterns for Characterization and Prediction of Financial Time Series Events

Richard J. Povinelli

Department of Electrical and Computer Engineering, Marquette University,
P.O. Box 1881, Milwaukee, WI 53201-1881, USA
Richard.Povinelli@Marquette.edu
http://povinelli.eece.mu.edu

Abstract. The novel Time Series Data Mining (TSDM) framework is applied to analyzing financial time series. The TSDM framework adapts and innovates data mining concepts to analyzing time series data. In particular, it creates a set of methods that reveal hidden temporal patterns that are characteristic and predictive of time series events. This contrasts with other time series analysis techniques, which typically characterize and predict all observations. The TSDM framework and concepts are reviewed, and the applicable TSDM method is discussed. Finally, the TSDM method is applied to time series generated by a basket of financial securities. The results show that statistically significant temporal patterns that are both characteristic and predictive of events in financial time series can be identified.

1 Introduction

The Time Series Data Mining (TSDM) framework [1-4] is applied to the prediction of financial time series. TSDM-based methods can successfully characterize and predict complex, nonperiodic, irregular, and chaotic time series. The TSDM methods overcome limitations (including stationarity and linearity requirements) of traditional time series analysis techniques by adapting data mining concepts for analyzing time series.

A time series is "a sequence of observed data, usually ordered in time" [5, p. 1]. Fig. 1 shows an example time series $X = \{x_t, t = 1, \ldots, N\}$, where t is a time index, and $N = 126$ is the number of observations. Time series analysis is fundamental to engineering, scientific, and business endeavors, such as the prediction of welding droplet releases and stock market price fluctuations [1, 2, 4].

This paper, which is divided into four sections, presents the results of applying the TSDM framework to the problem of finding a trading-edge, i.e., a small, but significant, advantage that allows greater than expected returns to be realized. The first section presents the problem and reviews other time series analysis techniques. The second section introduces the key TSDM concepts and method. The third section presents the prediction results. The fourth section discusses the results and proposes future work.

J.F. Roddick and K. Hornsby (Eds.): TSDM 2000, LNAI 2007, pp. 46-61, 2001.

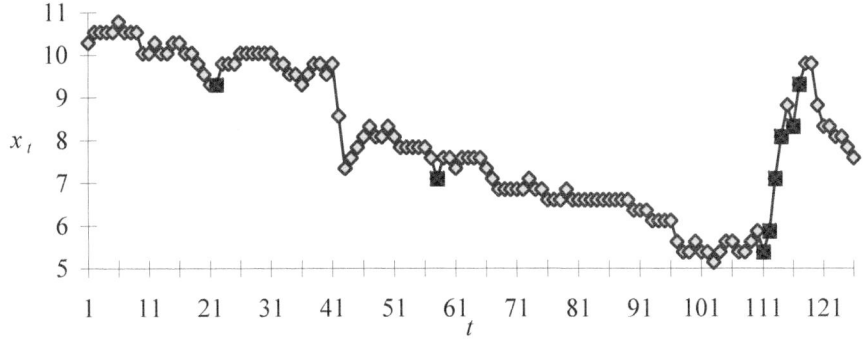

Fig. 1 – Stock Daily Open Price Time Series

1.1   Problem Statement

The predominant theory for describing the price behavior of a financial security is the efficient market hypothesis, which is explained using the expected return or fair game model [6, p. 210]. The expected value of a security is $E(P_{t+1}|\Phi_t) = (1 + E(r_{t+1}|\Phi_t))P_t$ [6, p. 210], where P_t is the price of a security at time t, r_{t+1} is the one-period percent rate of return for the security during period $t+1$, and Φ_t is the information assumed to be fully reflected in the security price at time t.

The three forms of the efficient market hypothesis are weak, semistrong, and strong. The weak form, which is relevant to this work, assumes Φ_t is all security-market information, such as historical sequence of price, rates of return, and trading volume data [6, p. 211].

The weak form of the efficient market hypothesis has been supported in the litera-ture [6, p. 213-215]. The efficient market hypothesis has been verified by showing that security price time series show no autocorrelation and are random according to the runs test. In addition, tests of trading rules have generally shown that the weak form of the efficient market hypothesis holds [6, p. 213-215].

The problem is to find a trading-edge, which is a small advantage that allows greater than expected returns to be realized. If the weak form of the efficient market hypothesis holds, the TSDM method should not be able to find hidden patterns that can be exploited to achieve such a trading-edge.

Fig. 1 illustrates the problem, where the horizontal axis represents time, and the vertical axis observations. The diamonds show the open price of a stock. The results of a successful prediction technique are illustrated by the black squares, which indi-cate buying opportunities. If the stock were purchased on those days and sold the next day, a greater than 5% return would be realized for each buy-sell sequence.

To summarize, the problem is to find hidden patterns that are, on average, charac-teristic and predictive of a larger than normal increase in the price of a stock and to use these hidden patterns in a trading strategy.

1.2   Review of Time Series Analysis Techniques

The analysis of financial time series has a long history. This review will briefly touch on some of the many time series analysis techniques that may be applied to predicting stock prices, including ARIMA, machine learning, genetic programming, neural network, and various data mining methods.

Some of the first applications of the traditional Box-Jenkins or Autoregressive Integrated Moving Average (ARIMA) method was to the analysis of the IBM stock time series [5]. The ARIMA techniques provide a comprehensive approach for analyzing stationary time series whose residuals are normal and independent [5]. For real-world time series such as stock market prices, the conditions of time series stationarity and residual normality and independence are not met. Another drawback of the ARIMA approach is its inability to identify complex hidden characteristics. This limitation occurs because of the goal of characterizing all time series observations.

For stock time series, the typical AR(1) model is $x_t = x_{t-1} + a_t$ [5, pp. 30-31], i.e., the expected next value in the time series is the current value. This model does not help in making trading decisions.

An example of applying machine learning techniques is provided by the work of Zemke, who uses a bagging approach to combine predictions made by an artificial neural network, a nearest neighbor method, and an evolved logic program to predict various stock indices [7]. Zemke is able to achieve an average daily excess return of 0.15% more than a random trading strategy.

Kaboudan uses a genetic programming approach to learn the nonlinear generating function to predict stock time series [8]. He develops a trading strategy that is tested against six stocks. Kaboudan is able to achieve an average daily excess return of 0.45% more than a naïve trading approach.

Berndt and Clifford [9], Keogh [10-12], Rosenstein and Cohen [13], and Guralnik et al. [14] are among those who have applied data mining concepts to finding patterns in time series. Data Mining [15, 16] is the analysis of data with the goal of uncovering hidden patterns. It encompasses a set of methods that automate the scientific discovery process. Its uniqueness is found in the types of problems addressed – those with large data sets and complex, hidden relationships. Data mining evolved from several fields, including machine learning, statistics, and database design [16]. It uses techniques such as clustering, association rules, visualization, decision trees, nonlinear regression, and probabilistic graphical dependency models to identify novel, hidden, and useful structures in large databases [15, 16].

Berndt and Clifford use a dynamic time warping technique taken from speech recognition. Their approach uses a dynamic programming method for aligning the time series and a predefined set of templates.

Rosenstein and Cohen [13] also use a predefined set of templates to match a time series generated from robot sensors. Instead of using the dynamic programming methods as in [9], they employ the time-delay embedding process to match their predefined templates.

Similarly, Keogh represents the templates using piecewise linear segmentations. "Local features such as peaks, troughs, and plateaus are defined using a prior distribution on expected deformations from a basic template" [10]. Keogh's approach uses a probabilistic method for matching the known templates to the time series data.

Guralnik et al. [14] have developed a language for describing temporal patterns (episodes) in sequence data. They have developed an efficient sequential pattern tree for identifying frequent episodes. Their work, like that of others discussed here, focuses on quickly finding patterns that match predefined templates.

The novel TSDM framework, initially introduced by Povinelli and Feng in [1], differs fundamentally from both data mining and other time series approaches. The TSDM framework differs from most time series analysis techniques by focusing on discovering hidden temporal patterns that are predictive of events, which are *important* occurrences, rather than trying to predict all observations. This allows the TSDM methods to predict nonstationary, nonperiodic, irregular time series, including chaotic deterministic time series. The TSDM methods are applicable to time series that appear stochastic, but occasionally (though not necessarily periodically) contain distinct, but possibly hidden, patterns that are characteristic of the desired events.

The data mining approaches advanced in [9-14] require *a priori* knowledge of the types of structures or temporal patterns to be discovered. These approaches represent temporal patterns as a set of templates. The use of predefined templates in [9-14] prevents the achievement of the basic data mining goal of discovering useful, novel, and hidden temporal patterns. The TSDM framework is not restricted by the use of predefined templates.

The novel TSDM framework creates a new structure for analyzing time series by adapting concepts from data mining [15, 16]; time series analysis [5, 17, 18]; genetic algorithms [19-21]; and chaos, nonlinear dynamics, and dynamical systems [22-25]. From data mining comes the focus on discovering hidden patterns. From time series analysis comes the theory for analyzing linear, stationary time series. In the end, the limitations of traditional time series analysis suggest the possibility of new methods. From genetic algorithms comes a robust and easily applied optimization method [19]. From the study of chaos, nonlinear dynamics, and dynamical systems comes the theoretical justification of the TSDM methods, specifically Takens' Theorem [26] and Sauer's extension [27].

2   Some Time Series Data Mining Concepts

Previous work [1, 2, 4] presented the TSDM framework. In this section, the fundamental TSDM concepts such as events, temporal patterns, event characterization function, temporal pattern cluster, time-delay embedding, phase space, augmented phase space, objective function, and optimization are defined and explained as is the TSDM method for identifying temporal pattern clusters.

The TSDM method discussed here discovers hidden temporal patterns (vectors of length Q) characteristic of events (important occurrences) by time-delay embedding [22, 25] an observe time series X into a reconstructed phase space, here simply called *phase space*. An event characterization function g is used to represent the eventness of a temporal pattern. An augmented phase space is formed by extending the phase space with g. The augmented phase space is searched for a temporal pattern cluster P that best characterizes the desired events. The temporal pattern clusters are then used to predict events in a testing time series.

2.1   Events, Temporal Pattern, and Temporal Pattern Cluster

In a time series, an event is an important occurrence. The definition of an event is dependent on the TSDM goal. For example, an event may be defined as the sharp rise or fall of a stock price. Let $X = \{x_t, t = 1,\ldots,126\}$ be the daily open price of a stock for a six-month period as illustrated by Fig. 1. The events, highlighted as squares in Fig. 1, are those days when the price increases more than 5%.

A temporal pattern is a hidden structure in a time series that is characteristic and predictive of events. The temporal pattern $\mathbf{p}$ is a real vector of length Q. The temporal pattern is represented as a point in a Q dimensional real metric space, i.e., $\mathbf{p} \in \mathbb{R}^Q$.

Because a temporal pattern may not perfectly match the time series observations that precede events, a temporal pattern cluster is defined as the set of all points within δ of the temporal pattern. The temporal pattern cluster $P = \{ a \in \mathbb{R}^Q : d(\mathbf{p}, a) \le \delta \}$, where d is the distance or metric defined on the space. This defines a hypersphere of dimension Q, radius δ, and center $\mathbf{p}$.

The observations $\{x_{t-(Q-1)\tau}, \ldots, x_{t-2\tau}, x_{t-\tau}, x_t\}$ form a sequence that can be compared to a temporal pattern, where x_t represents the current observation, and $x_{t-(Q-1)\tau}, \ldots, x_{t-2\tau}, x_{t-\tau}$ past observations. Let $\tau > 0$ be a positive integer. If t represents the present time index, then $t - \tau$ is a time index in the past, and $t + \tau$ is a time index in the future. Using this notation, time is partitioned into three categories: past, present, and future. Temporal patterns and events are placed into different time categories. Temporal patterns occur in the past and complete in the present. Events occur in the future.

2.2   Phase Space and Time-Delay Embedding

A reconstructed phase space [22] is a Q-dimensional metric space into which a time series is embedded. Takens showed that if Q is large enough, the phase space is homeomorphic to the state space that generated the time series [26]. The time-delayed embedding of a time series maps a set of Q time series observations taken from X onto $\mathbf{x}_t$, where $\mathbf{x}_t$ is a vector or point in the phase space. Specifically, $\mathbf{x}_t = (x_{t-(Q-1)\tau}, \ldots, x_{t-2\tau}, x_{t-\tau}, x_t)$.

2.3   Event Characterization Function

To link a temporal pattern (past and present) with an event (future) the event characterization function $g(t)$ is introduced. The event characterization function represents the value of future "eventness" for the current time index. It is, to use an analogy, a measure of how much gold is at the end of the rainbow (temporal pattern). The event characterization function is defined *a priori* and is created to address the specific TSDM goal. The event characterization function is defined such that its value at *t* correlates highly with the occurrence of an event at some specified time in the future, i.e., the event characterization function is causal when applying the TSDM method to prediction problems. Non-causal event characterization functions are useful when applying the TSDM method to system identification problems.

In Fig. 1, the goal is to decide if the stock should be purchased today and sold tomorrow. The event characterization function that achieves this goal is $g(t) = (x_{t+1} - x_t)/x_t$, which assigns the percentage change in the stock price for the next day to the current time index. Alternatively the time series maybe filtered, thereby simplifying the event characterization function.

2.4   Augmented Phase Space

The concept of an augmented phase space follows from the definitions of the event characterization function and the phase space. The augmented phase space is a $Q+1$ dimensional space formed by extending the phase space with $g(\cdot)$ as the extra dimension. Every augmented phase space point is a vector $<\mathbf{x}_t, g(t) >\in \mathbb{R}^{Q+1}$.

Fig. 2, a stem-and-leaf plot, shows the augmented phase space for the daily return time series generated from the open price time series illustrated in Fig. 1. The height of the leaf represents the significance of $g(\cdot)$ for that time index.

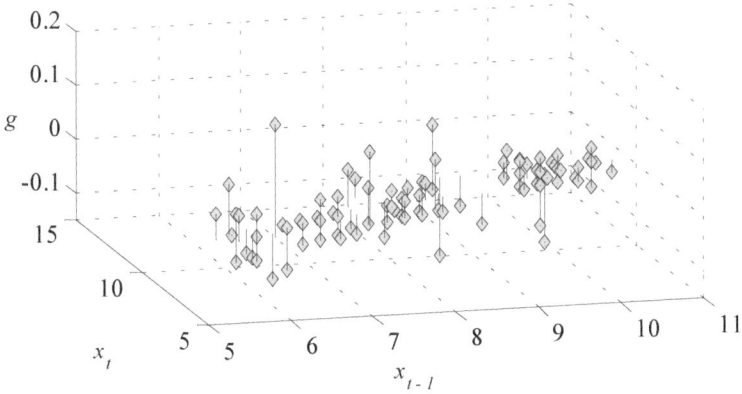

Fig. 2 – Stock Daily Return Augmented Phase Space

2.5   Objective Function

The TSDM objective function represents the efficacy of a temporal pattern cluster to characterize events. The objective function *f* maps a temporal pattern cluster *P* onto the real line, which provides an ordering to temporal pattern clusters according to their ability to characterize events. The objective function is constructed in such a manner that its optimizer P^* meets the TSDM goal.

The form of the objective functions is application dependent, and several different objective functions may achieve the same goal. Before presenting an example objective function, several definitions are required.

The index set $\Lambda = \{t : t = (Q-1)\tau + 1, \ldots, N\}$, where $(Q-1)\tau$ is the largest embedding time-delay, and N is the number of observations in the time series, is the set of all time indices t of phase space points. The index set M is the set of all time indices t when x_t is within the temporal pattern cluster, i.e. $M = \{t : x_t \in P, t \in \Lambda\}$.

The average value of g, also called the average eventness, of the phase space points within the temporal pattern cluster P is

$$\mu_M = \frac{1}{c(M)} \sum_{t \in M} g(t),$$

where $c(M)$ is the cardinality of M.

The following objective function orders temporal pattern clusters according to their ability to characterize time series observations with high eventness and characterize at least a minimum number of events. The objective function

$$f(P) = \begin{cases} \mu_M & \text{if } c(M)/c(\Lambda) \geq \beta \\ (\mu_M - g_0)\dfrac{c(M)}{\beta c(\Lambda)} + g_0 & \text{otherwise} \end{cases}, \quad (1)$$

where β is the desired minimum percentage cardinality of the temporal pattern cluster, and g_0 is the minimum eventness of the phase space points, i.e. $g_0 = \min\{g(t) : t \in \Lambda\}$.

The parameter β in the linear barrier function in (1) is chosen so that $c(M)$ is non-trivial, i.e., the neighborhood around p includes some percentage of the total phase space points. If $\beta = 0$, then $c(M) = 1$ or $g(i) = g(j) \; \forall i, j \in M$, i.e., the eventness value of all points in the temporal pattern cluster are identical. If $\beta = 0$, the temporal pattern cluster will be maximal when it contains only one point in the phase space – the point with the highest eventness. If there are many points with the highest eventness, the optimal temporal pattern cluster may contain several of these points. When $\beta = 0$, (1) is still defined, because $c(M)/c(\Lambda) \geq 0$ is always true.

2.6   Optimization

The key to the TSDM framework is finding optimal temporal pattern clusters that characterize and predict events. Thus, an optimization algorithm to maximize $f(P)$ over p and δ is necessary. A modified simple GA (sGA) [19] composed of a Monte Carlo initialization, roulette selection, and random locus crossover is used for finding P^*. The Monte Carlo search generates the initial population for the sGA. Although a mutation operator is typically incorporated into an sGA, it is not used for discovering the results presented in this paper. The sGA uses a binary chromosome with gene lengths of six and single individual elitism. The stopping criterion for the GA is convergence of all fitness values. The population size is 30 and the Monte Carlo search size is 300. A hashing technique is employed to improve computational performance [28].

2.7   Time Series Data Mining Method

The first step in applying the TSDM method is to define the TSDM goal, which is specific to each application, but may be stated generally as follows. Given an observed time series $X = \{x_t, t = 1, \ldots, N\}$, the goal is to find hidden temporal patterns that are characteristic of events in X, where events are specified in the context of the problem. Likewise, given a testing time series $Y = \{x_t, t = R, \ldots, S\}$ $N < R < S$, the goal is to use the hidden temporal patterns discovered in X to predict events in Y.

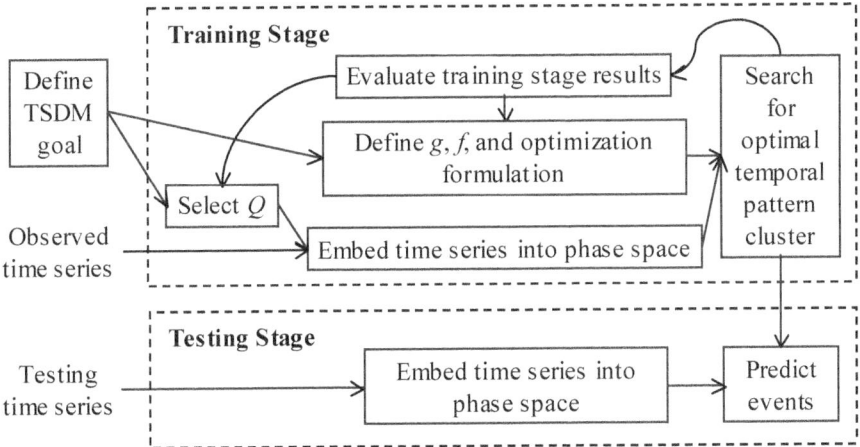

Fig. 3 – Block Diagram of TSDM Method

Given a TSDM goal, an observed time series to be characterized, and a testing time series to be predicted, the steps in the TSDM method are:
Training Stage (Batch Process)
1. Frame the TSDM goal in terms of the event characterization function, objective function, and optimization formulation.
    a. Define the event characterization function g.
    b. Define the objective function f.
    c. Define the optimization formulation, including the independent variables over which the value of the objective function will be optimized and the constraints on the objective function.
2. Determine Q, i.e., the dimension of the phase space and the length of the temporal pattern.
3. Transform the observed time series into the phase space using the time-delayed embedding process.

4. Associate with each time index in the phase space, an eventness represented by the event characterization function. Form the augmented phase space.
5. In the augmented phase space, search for the optimal temporal pattern cluster, which best characterizes the events.
6. Evaluate training stage results. Repeat training stage as necessary.

Testing Stage (Real Time or Batch Process)
1. Embed the testing time series into the phase space.
2. Use the optimal temporal pattern cluster for predicting events.
3. Evaluate testing stage results.

3   Financial Applications of Time Series Data Mining

This section presents significant results found by applying the Time Series Data Mining (TSDM) method to a basket of financial time series. The time series are created by the dynamic interaction of millions of investors buying and selling securities through a secondary equity market such as the New York Stock Exchange (NYSE) or National Association of Securities Dealers Automated Quotation (NASDAQ) market [6]. The times series are measurements of the activity of a security, specifically a stock.

The goal is to find a trading-edge, a small advantage that allows greater than expected returns to be realized. If the weak form of the efficient market hypothesis holds, the TSDM method should not be able to find temporal patterns that can be exploited to achieve such a trading-edge. The TSDM goal is to find temporal pattern clusters that are, on average, characteristic and predictive of a larger than normal increase in the price of a stock.

Two sets of time series are analyzed. The first set of time series are the inter-day returns for the 30 Dow Jones Industrial Average (DJIA) components from January 2, 1990 through March 8, 1991. This time period allows for approximately 200 testing stages. The inter-day return $r_t = (o_{t+1} - o_t)/o_t$, where o_t is the daily open price, which is the price of the first trade. Detailed results for this set of time series are provided.

The second set of time series are the intra-day returns for the 30 DJIA components from October 16, 1998 through December 22, 1999. Again, this time period allows for approximately 200 testing stages. The intra-day return $r_t = (c_t - o_t)/o_t$, where o_t is the daily open price and c_t is the daily closing price, which is the price of the last trade. Summary results are provided for this set of time series.

Fig. 4 illustrates the DJIA during the first time period.

3.1   Training Stage

The TSDM method, illustrated in Fig. 4, is applied 198 times to each of the DJIA component time series for a total of 5,940 training stages for each set of time series. The 198 observed time series are formed from a moving window of length 100. The testing time series is a single observation. The parameters of the method are:

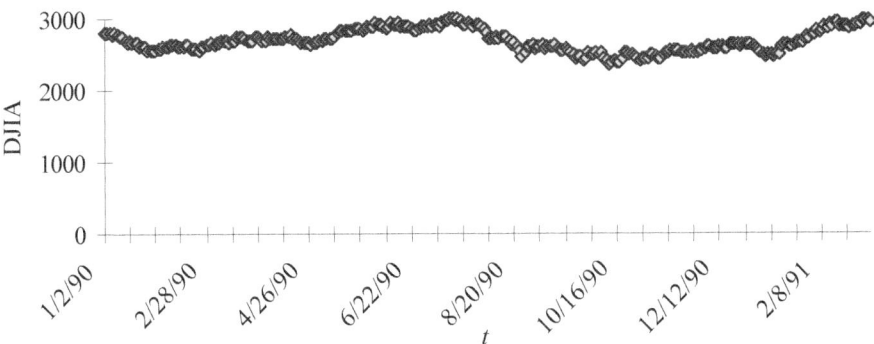

Fig. 4 – DJIA Daily Open Price Time Series

- The event characterization function $g(t) = x_{t+1}$, which allows for one-step-ahead characterization and prediction.
- The objective function (1) with a $\beta = 0.05$ is used.
- The optimization formulation is max $f(P)$.
- The dimension of the phase space $Q = 2$.

The statistical training results for each DJIA component are presented in Table 1. Of the 5,940 training processes, the cluster mean eventness (μ_M) was greater than total mean eventness (μ_X) every time. For 69% of the temporal pattern clusters, the probability of a Type I error (α) was less than 5% based on the independent means statistical test.

Table 1 – DJIA Component Results, January 2, 1990, through March 8, 1991 (Observed)

Ticker	$\mu_M > \mu_X$	$\alpha \leq 0.05$
AA	100%	82%
ALD	100%	72%
AXP	100%	71%
BA	100%	70%
CAT	100%	79%
CHV	100%	54%
DD	100%	42%
DIS	100%	83%
EK	100%	55%
GE	100%	66%
GM	100%	73%
GT	100%	62%
HWP	100%	55%
IBM	100%	67%
IP	100%	80%

Ticker	$\mu_M > \mu_X$	$\alpha \le 0.05$
JNJ	100%	89%
JPM	100%	90%
KO	100%	67%
MCD	100%	62%
MMM	100%	57%
MO	100%	65%
MRK	100%	59%
PG	100%	76%
S	100%	59%
T	100%	66%
TRV	100%	78%
UK	100%	36%
UTX	100%	94%
WMT	100%	73%
XON	100%	75%
Combined	100%	69%

3.2   Testing Stage

Using the 5,940 training processes, 471 events are predicted for the January 2, 1990 through March 8, 1991 time series. The statistical prediction results for each DJIA component are presented in Table 2. The cluster mean eventness (μ_M) was greater than the non-cluster mean eventness ($\mu_{\tilde{M}}$) 20 out of 30 times or 67% of the time. For 16.7% of the temporal pattern clusters, the probability of a Type I error was less than 5% based on the independent means statistical test. These low rates of statistical significance at the 5% α level are typical for predictions of financial time series [1, 2].

Table 2 – DJIA Component Results, January 2, 1990, through March 8, 1991 (Testing)

Ticker	$c(M)$	μ_M	σ_M	$c(M)$	$\mu_{\tilde{M}}$	$\sigma_{M'}$	α_m
AA	16	0.569%	1.652%	182	-0.013%	1.620%	1.78×10^{-1}
ALD	14	0.438%	1.428%	184	-0.102%	1.851%	1.83×10^{-1}
AXP	14	0.027%	2.058%	184	-0.023%	2.610%	9.32×10^{-1}
BA	13	0.080%	2.044%	185	-0.030%	2.181%	8.52×10^{-1}
CAT	26	-0.003%	1.817%	172	-0.098%	2.127%	8.08×10^{-1}
CHV	16	0.057%	1.572%	182	0.061%	1.200%	9.92×10^{-1}
DD	16	0.526%	1.946%	182	-0.045%	1.635%	2.55×10^{-1}
DIS	20	-0.024%	1.488%	178	0.069%	2.069%	8.00×10^{-1}
EK	14	-0.045%	1.879%	184	0.074%	1.998%	8.20×10^{-1}
GE	16	0.094%	1.410%	182	0.000%	1.881%	8.04×10^{-1}
GM	16	0.671%	2.090%	182	-0.149%	1.863%	1.29×10^{-1}
GT	20	-0.962%	2.034%	178	-0.066%	2.549%	6.93×10^{-2}

Ticker	$c(M)$	μ_M	σ_M	$c(\tilde{M})$	$\mu_{\tilde{M}}$	$\sigma_{M'}$	α_m
HWP	13	-0.779%	1.881%	185	0.116%	2.664%	1.08×10^{-1}
IBM	16	-1.079%	1.785%	182	0.175%	1.460%	6.32×10^{-3}
IP	16	1.197%	2.525%	182	0.025%	1.587%	6.80×10^{-2}
JNJ	13	0.665%	1.444%	185	0.160%	1.551%	2.25×10^{-1}
JPM	11	1.420%	1.878%	187	0.040%	1.985%	1.82×10^{-2}
KO	11	1.794%	3.396%	187	0.008%	1.807%	8.36×10^{-2}
MCD	13	0.367%	1.753%	185	-0.013%	1.977%	4.54×10^{-1}
MMM	16	0.238%	1.044%	182	0.043%	1.258%	4.82×10^{-1}
MO	17	0.038%	1.820%	181	0.251%	1.641%	6.42×10^{-1}
MRK	19	0.669%	1.163%	179	0.073%	1.580%	4.11×10^{-2}
PG	13	0.174%	1.615%	185	0.047%	1.707%	7.85×10^{-1}
S	14	1.449%	2.677%	184	-0.157%	1.938%	2.77×10^{-2}
T	11	1.307%	1.797%	187	-0.193%	1.645%	6.88×10^{-3}
TRV	21	1.531%	2.449%	177	-0.147%	2.617%	3.21×10^{-3}
UK	14	-0.449%	2.263%	184	0.041%	1.900%	4.30×10^{-1}
UTX	14	-0.289%	1.979%	184	-0.028%	1.828%	6.33×10^{-1}
WMT	18	0.658%	1.950%	180	0.120%	2.458%	2.77×10^{-1}
XON	20	0.077%	1.398%	178	0.090%	1.263%	9.68×10^{-1}
All	471	0.313%	1.970%	5,469	0.011%	1.919%	1.38×10^{-3}

For the combined results – using all predictions – the mean cluster eventness is greater than the non-cluster mean eventness. It also is statistically significant to the 0.005α level according to the independent means test.

The best way to understand the effectiveness of the TSDM method when applied to financial time series is to show the trading results that can be achieved by applying the temporal pattern clusters discovered above. An initial investment is made as follows: If a temporal pattern cluster from any of the stocks in the portfolio predicts a high eventness, the initial investment is made in that stock for one day. If there are temporal pattern clusters for several stocks that indicate high eventness, the initial investment is split equally among the stocks. If there are no temporal pattern clusters indicating high eventness, then the initial investment is invested in a money market account with an assumed 5% annual rate of return. The training process is rerun using the new 100 most recent observation window. The following day, the initial investment principal plus return is invested according to the same rules. The process is repeated for the remaining investment period.

The results for the investment period of May 29, 1990 through March 8, 1991 are shown in Table 3. This period is shorter than the total time frame (January 1, 1990 through March 8, 1991) because the first part of the time series is used only for training. The return of the DJIA also is given, which is slightly different from the buy and hold strategy for all DJIA components because the DJIA has a non-equal weighting among its components.

58   Richard J. Povinelli

Table 3 – Trading Results, May 29, 1990 through March 8, 1991

Portfolio	Investment Method	Return	Annualized Return
All DJIA components	Temporal Pattern Cluster	30.98%	41.18%
DJIA	Buy and Hold	2.95%	3.79%
All DJIA components	Not in Temporal Pattern Cluster	0.35%	0.45%
All DJIA components	Buy and Hold	3.34%	4.29%

The results for the investment period of March 15, 1999 through December 22, 1999 are shown in Table 4. Again, this period is shorter than the total time frame (October 16, 1998 through December 22, 1999) because the first part of the time series is used only for training. The return of the DJIA varies significantly from the buy and hold strategy for all DJIA components not only because the DJIA has a non-equal weighting among its components, but more importantly because intra-day return time series are used. The results for this set of time series is less significant than the previous with an $\alpha = 0.12$.

Table 4 – Trading Results, March 15, 1999 through December 22, 1999

Portfolio	Investment Method	Return	Annualized Return
All DJIA components	Temporal Pattern Cluster	22.70%	29.88%
DJIA	Buy and Hold	13.39%	17.42%
All DJIA components	Not in Temporal Pattern Cluster	-10.92%	-13.74%
All DJIA components	Buy and Hold	-8.26%	-10.43%

An initial investment of $10,000 made on May 29, 1990 in the 30 DJIA component stocks using the TSDM method would have grown to $13,098 at the end of March 8, 1991. The maximum draw down, the largest loss during the investment period, is 9.65%. An initial investment of $10,000 made on March 15, 1999 using the TSDM method would have grown to $12,700 at the end of December 22, 1999 with a maximum draw down of 10.2%. One caveat to this result is that it ignores trading costs [29]. The trading cost is a percentage of the amount invested and includes both the buying and selling transaction costs along with the spread between the bid and ask, where the bid is the offer price for buying and the ask is the offer price for selling. The trading cost in percentage terms would need to be kept in the 0.02% range. This level of trading cost would require investments in the $500,000 to $1,000,000 range and access to trading systems that execute in between the bid and ask prices or have spreads of 1/16th or less.

4   Conclusions and Future Work

Through the novel Time Series Data Mining (TSDM) framework and its associated method, this paper has made an original contribution to the fields of time series analysis and data mining. The key TSDM concepts of event, event characterization function, temporal pattern, temporal pattern cluster, time-delay embedding, phase space, augmented phase space, objective function, and optimization were reviewed, setting up the framework from which to develop TSDM methods.

The TSDM method was successfully applied to characterizing and predicting complex, nonstationary time series events from the financial domain. In the financial domain, it was able to generate a trading-edge.

Future research efforts will involve the direct comparison over the same time periods of the TSDM method present here with the techniques proposed by Zemke [7] and Kaboudan [8]. Additional comparisons with Hidden Markov Model techniques also will be investigated. A detailed study of the risk-return characteristics of these various methods will be undertaken.

Additionally, new time series predictability metrics will be created that specifically address the event nature of the TSDM framework. This research direction will study the characteristics of time series that allow for the successful application of the TSDM framework.

References

[1]   R. J. Povinelli and X. Feng, "Temporal Pattern Identification of Time Series Data using Pattern Wavelets and Genetic Algorithms," proceedings of Artificial Neural Networks in Engineering, St. Louis, Missouri, 1998, pp. 691-696.

[2]   R. J. Povinelli and X. Feng, "Data Mining of Multiple Nonstationary Time Series," proceedings of Artificial Neural Networks in Engineering, St. Louis, Missouri, 1999, pp. 511-516.

[3]   R. J. Povinelli, "Using Genetic Algorithms to Find Temporal Patterns Indicative of Time Series Events," proceedings of Genetic and Evolutionary Computation Conference (GECCO-2000) Workshop Program, Las Vegas, Nevada, 2000, pp. 80-84.

[4]   R. J. Povinelli, *Time Series Data Mining: Identifying Temporal Patterns for Characterization and Prediction of Time Series Events*, Ph.D. Dissertation, Marquette University, 1999.

[5]   S. M. Pandit and S.-M. Wu, *Time series and system analysis, with applications*. New York: Wiley, 1983.

[6]   F. K. Reilly and K. C. Brown, *Investment analysis and portfolio management*, 5th ed. Fort Worth, Texas: Dryden Press, 1997.

[7]   S. Zemke, "Bagging imperfect predictors," proceedings of Artificial Neural Networks in Engineering, St. Louis, Missouri, 1999, pp. 1067-1072.

[8]   M. Kaboudan, "Genetic Programming Prediction of Stock Prices," *Computational Economics*, to appear.

[9]   D. J. Berndt and J. Clifford, "Finding Patterns in Time Series: A Dynamic Programming Approach," in *Advances in knowledge discovery and data mining*, U. M. Fayyad, G. Piatetsky-Shapiro, P. Smyth, and R. Uthursamy, Eds. Menlo Park, California: AAAI Press, 1996, pp. 229-248.

[10]   E. Keogh and P. Smyth, "A Probabilistic Approach to Fast Pattern Matching in Time Series Databases," proceedings of Third International Conference on Knowledge Discovery and Data Mining, Newport Beach, California, 1997.

[11]   E. Keogh, "A Fast and Robust Method for Pattern Matching in Time Series Databases," proceedings of 9th International Conference on Tools with Artificial Intelligence (TAI '97), 1997.

[12]   E. J. Keogh and M. J. Pazzani, "An enhanced representation of time series which allows fast and accurate classification, clustering and relevance feedback," proceedings of AAAI Workshop on Predicting the Future: AI Approaches to Time-Series Analysis, Madison, Wisconsin, 1998.

[13]   M. T. Rosenstein and P. R. Cohen, "Continuous Categories For a Mobile Robot," proceedings of Sixteenth National Conference on Artificial Intelligence, 1999.

[14]   V. Guralnik, D. Wijesekera, and J. Srivastava, "Pattern Directed Mining of Sequence Data," proceedings of International Conference on Knowledge Discovery and Data Mining, New York, NY, 1998, pp. 51-57.

[15]   U. M. Fayyad, G. Piatetsky-Shapiro, P. Smyth, and R. Uthursamy, *Advances in knowledge discovery and data mining*. Menlo Park, California: AAAI Press, 1996.

[16]   S. M. Weiss and N. Indurkhya, *Predictive data mining: a practical guide*. San Francisco: Morgan Kaufmann, 1998.

[17]   G. E. P. Box and G. M. Jenkins, *Time series analysis: forecasting and control*, Rev. ed. San Francisco: Holden-Day, 1976.

[18]   B. L. Bowerman and R. T. O'Connell, *Forecasting and time series: an applied approach*, 3rd ed. Belmont, California: Duxbury Press, 1993.

[19]   D. E. Goldberg, *Genetic algorithms in search, optimization, and machine learning*. Reading, Massachusetts: Addison-Wesley, 1989.

[20]   R. J. Povinelli and X. Feng, "Improving Genetic Algorithms Performance By Hashing Fitness Values," proceedings of Artificial Neural Networks in Engineering, St. Louis, Missouri, 1999, pp. 399-404.

[21]   J. H. Holland, *Adaptation in natural and artificial systems: an introductory analysis with applications to biology, control, and artificial intelligence*, 1st MIT Press ed. Cambridge, Massachusetts: MIT Press, 1992.

[22]   H. D. I. Abarbanel, *Analysis of observed chaotic data*. New York: Springer, 1996.

[23]   A. J. Crilly, R. A. Earnshaw, and H. Jones, *Applications of fractals and chaos*. Berlin: Springer, 1993.

[24]   N. B. Tufillaro, T. Abbott, and J. Reilly, *An experimental approach to nonlinear dynamics and chaos*. Redwood City, California: Addison-Wesley, 1992.

[25]   J. Iwanski and E. Bradley, "Recurrence plot analysis: To embed or not to embed?," Chaos, vol. 8, pp. 861-871, 1998.

[26]   F. Takens, "Detecting strange attractors in turbulence," proceedings of Dynamical Systems and Turbulence, Warwick, 1980, pp. 366-381.

[27]   T. Sauer, J. A. Yorke, and M. Casdagli, "Embedology," *Journal of Statistical Physics*, vol. 65, pp. 579-616, 1991.

[28]   R. J. Povinelli, "Improving Computational Performance of Genetic Algorithms: A Comparison of Techniques," proceedings of Genetic and Evolutionary Computation Conference (GECCO-2000) Late Breaking Papers, Las Vegas, Nevada, 2000, pp. 297-302.

[29]   J. D. Freeman, "Behind the smoke and mirrors: Gauging the integrity of investment simulations," *Financial Analysts Journal*, vol. 48, pp. 26-31, 1992.

Value Range Queries on Earth Science Data via Histogram Clustering

Ruixin Yang, Kwang-Su Yang, Menas Kafatos, and X. Sean Wang

Center for Earth Observing & Space Research
George Mason University
Fairfax, VA 22030, U.S.A.
{ryang, kyang, mkafatos, xywang}@gmu.edu

Abstract. Remote sensing data as well as ground-based and model output data about the Earth system can be very large in volume. On the other hand, in order to use the data efficiently, scientists need to search for data based on not only metadata but also actual data values. To answer value range queries by scanning very large volumes of data is obviously unrealistic. This article studies a clustering technique on histograms of data values on predefined cells to index the cells. Through this index system, the so-called statistical range queries can be answered quickly and approximately together with an accuracy assessment. Examples of using this technique for Earth science data sets are given in this article.

1   Introduction

In the last two decades, great progress has been made in Earth observing from space. Unprecedented observations of the Earth have provided a very large amount of data. In the next decade, the Earth Observing System (EOS) and other Earth observing platforms will produce massive data products with total rates of more than 1.5 Terabytes/day[1]. In order to use the data more efficiently, a fast way to search for user interested data is important. Moreover, scientists need to search data based on not only metadata but also actual data values. For example, a user may be interested in the regions over which a given parameter has a value in a certain range. In this article, we study efficient methods for such value range queries on gridded Earth science data.

A value range query on one data set may not be of much value by itself. However, it is a basic step for answering more sophisticated queries. For example, application scenarios of queries on Earth science data are recently discussed by a group of people under NASA's Earth Science Information Partners project[2]. Almost in all scenarios[3], value range queries are needed to obtain query results for the applications. For very large data sets, it is obviously unrealistic to answer value range queries by scanning very large volumes of data. In this article, we use an approach to approximately answer value range queries.

Approximate answers to database queries by using aggregation have been studied quite extensively in the literature. The concept of data cube[4] was introduced to store the aggregation values. Online aggregation processes have also

J.F. Roddick and K. Hornsby (Eds.): TSDM 2000, LNAI 2007, pp. 62–76, 2001.

been studied[5, 6] in which the aggregation values are computed by sampling method. Histograms, as one kind of aggregations, have been used in databases for efficient estimation of query result size. The histograms contain the contents of relations and are used to approximately answer range queries, or in other words, to give a selectivity estimation[7]. Poosala and Ganti[8] even use histograms to summarize the aggregation values (contents of data cube) to approximately answer aggregation queries. Carr and Olsen[9] sort histograms in plots to reduce the time needed to make decisions on whether further studies are justified for the plots. Here, we cluster histograms and use the clusters as indices to the database.

Clustering techniques have been used quite extensively in various applications[10, 11]. This technique is mostly used for data exploration such as pattern classification, image segmentation, etc. The focus of this kind of work is to explore relations among different groups of data (clusters), that is, data mining. Recently, Jain et al. [10] review clustering methods extensively from a perspective of a statistical pattern recognition, providing a taxonomy of clustering techniques and their applications such as image segmentation, object recognition, information retrieval, and data mining. Puzicha et al. [12] introduce clustering method for distributional and histogram data. The focus is on computing optimal clustering solutions by the annealed maximum a posteriori estimation based on the Bayesian framework.

In this work, we extend and apply clustering methods to gridded Earth system science data products. Our method makes it possible for the system to quickly answer the queries which return the regions on which more than a certain percentage of the underlying full resolution data falls in the given value range. The use of such queries will help the users to locate the regions of interests, which are often smaller than the whole region on which data are available. This will reduce the volume of the data that users need to download, order, or purchase.

The rest of the paper is organized as follows. In the next section, Section 2, the clustering method we propose is described. Examples for Earth science data sets are given in Section 3. Finally, discussion and future potential work are presented in Section 4.

2   Methods

The purpose of a value range search is to find spatial/temporal regions on which parameter values fall in certain ranges. The speed for answering such a query and the accuracy of the result may affect each other inversely. The goal of this work is to rapidly answer value range queries with reasonable accuracies. Instead of finding *all* regions satisfying the conditions, we want to **quickly** find not less than a certain percentage (say, 95%) of the total qualified regions. This kind of approximate queries are termed statistical range queries[13].

A data pyramid model[13] has been proposed to answer statistical range queries. In this data model, the problem space is divided into cells at different levels. In each level, all cells are disjointed and exhaustively cover the whole study space. Each higher level cell covers a space which is the union of a number of lower level cells. Each low level cell is associated with one and only one upper level cell. The construction of a data pyramid based on this model is bottom-up. Values associated with a level are computed from the values of the immediately lower layer. To answer a statistical range query, the procedure, on the other hand, is a drilling-down procedure[13]. That is, the procedure is to search the highest level first to find regions in the immediate lower resolution, then, to drill one layer down to search within the selected regions only. By limiting the lower level searches within the selected regions only, one may save a significant amount of computing time. However, this drill-down algorithm could not guarantee the specific accuracy given in a statistical query when multiple (more than two) levels are involved. As such, it suffers inherent limitations for multiple resolution pyramids.

In this work, we divide the problem domain following the same guideline. However, instead of building a data pyramid, we only create one low resolution layer above the original high resolution layer. The high resolution data values associated with one low resolution cell are used to create a histogram for that cell. Then a clustering method is used to cluster all histograms and the clustering result is used to index low resolution cells. When a query is answered, whether to include a cell in the query result or not depends on the associated index value of the cell. Cells of the same index would be taken into account in a group only. Computation of probabilities (counts) is performed on a few representative histograms instead of the set of all individual histograms associated with all cells. In this way, we can answer statistical range queries not only with a high speed but also with a correctly assessed accuracy.

2.1 Histograms

For a given data set, data coverage dimensions should be known in space, time, and parameter domain. Then, according to the scientific problem, we can define low resolution cells in certain dimensions. In this work, we define cells only on 2-D spatial regions, that is, in longitude and latitude dimensions. The spatial domain with the original spatial resolution of the given data sets are divided into low resolution cells. Each of the high level (low resolution) cells contains a fixed number of low level cells.

Associated with each low resolution cell, there is a data value histogram. The histogram is created by binning data values in high resolution cells covered by the corresponding low resolution cells. To get a histogram with high accuracy and a fixed number of bins, the best way is to use adaptive bin sizes according to the minimum and maximum values associated with this histogram. However, we will use overall histogram similarity to cluster the cells. To compute similarities or distances between two histograms, we need to have all the histograms cover the same value range with the same bin sizes. Therefore, we choose the global minimum and maximum values to define an universal histogram frame.

2.2   Clustering Method

We use the hierarchical clustering method[11] applied to histograms to cluster cells. Each histogram could be treated as a one-dimensional array, or a vector. The distance between two vectors can be measured with a specific metric. Here, the metric used is the Manhattan (city block) metric[14] in which a distance between two vectors is measured by summing up the absolute difference of the corresponding components of the two vectors. After the distances between all pairs of vectors are computed, the group-average, agglomerative hierarchical method is used to create a dendrogram. Based on the dendrogram, the number of clusters is chosen. There is no practical optimal method for picking a particular number of clusters. Therefore, we tested several numbers of clusters in a given problem and studied the effect of different numbers of clusters.

It should be pointed out that the clustering methods used here is only for prototype purpose. In an operational system, different clustering method should be considered for treating large data sets such as those developed recently[15, 16, 17].

2.3   Query Execution

For the clusters of cells, each cell belongs to one and only one cluster. The individual histograms in a cluster are summed up to form a representative histogram of the group. When a statistical range query is issued, the system then computes the count of bottom cells with data values in the particular range. The clusters are ordered according to their counts in the representative histograms. For any defined accuracy, a certain number of groups are included in the search. The answer on the high resolution level could be achieved by scanning cells in the selected clusters only. Therefore, the searching time is reduced by searching only a fraction of the total number of cells. The accuracy is defined as the ratio of the selected number of low level cells to the total number of cells in the given value range. As more and more groups are included in the search, higher and higher accuracy is, obviously, reached. The advantage of this clustering method is that the clusters are ordered. Therefore, clusters of high contribution to the result are included first. Consequently, the accuracy increases quickly with the first few clusters.

2.4   Software

A commercial software, Splus, is chosen for constructing the clusters and answering queries. The major tasks for creating the clusters are: reading data values, constructing histograms, computing the distances between all pairs of histograms, establishing a clustering tree, and clustering high level cells. Splus provides built-in functions to do most of above computations. With a statistical range query, the program should do the following steps: computing the counts of values in the given range for all clusters, finding the order of clusters to be searched, determining how many clusters should be searched according to the given required accuracy, and then scanning the selected high level cells to find cells in low level on which the parameter values are inside the specific range. Splus is used for all the steps. Therefore, we have only one Splus program to execute to obtain the results. Of course, there are some control parameters as input to the program to make some necessary adjustments. In special cases, original data files may not be in the format suitable for the Splus program. In those cases, other programs are used to reformat the original data.

3   Experiments

3.1   Data and Data Pyramid

To evaluate the clustering method described above, we take a gridded global monthly Normalized Difference Vegetation Index (NDVI) data set[18] as an example. NDVI values are derived from the NOAA polar orbiter AVHRR instruments by normalizing the difference of infrared and red reflectance measurements[19]. The NDVI values provide a strong vegetation signal and are used widely for classification of vegetation. In the classification process, finding areas for which NDVI values fall in a certain range is useful. So are the statistical range queries. The NDVI data we are using are of 1×1 degree spatial resolution in both longitude and latitude. The spatial coverage of the data set is global land. Since we divide only the spatial domain into cells, a fixed time, August, 1981, was randomly chosen.

   We take the original resolution areas of 1×1 degree cells as the high resolution (low level) cells and use 5×5 degree cells on the high level (low resolution). For each high level cell, there are up to 25 associated cells in the low level. The 25 NDVI data values on each low level cell are mapped into a fixed bin frame to form a histogram. In this case, we use 22 bins of width 0.05 covering the value range from -0.1 to 1.0. The mapping result is a histogram for each high level cell. Figure 1 shows some of the histograms. Since the NDVI data cover only global land, there are many empty cells. The empty cells are dropped when the cluster indices are created. Even so, some histograms are composed of less than 25 values because the corresponding high level cells cover both land and water.

3.2   Clustering Result

The result of the clustering process is a dendrogram (not shown) and the diagram is used as a reference for picking up the number of clusters. We first separate the high level cells into six groups to demonstrate the efficiency of this method. Then, we study the influence of the group number chosen for a given data set. Figure 2 displays several histograms for each group. Each row represents histograms from one group. It is obvious that the shapes of histograms in any one group are similar. In contrast, the distribution of histograms in Figure 1 is quite random. The result is reasonable and shows the power of clustering. For each group, we sum up the values of all histograms and obtain a representative histogram.

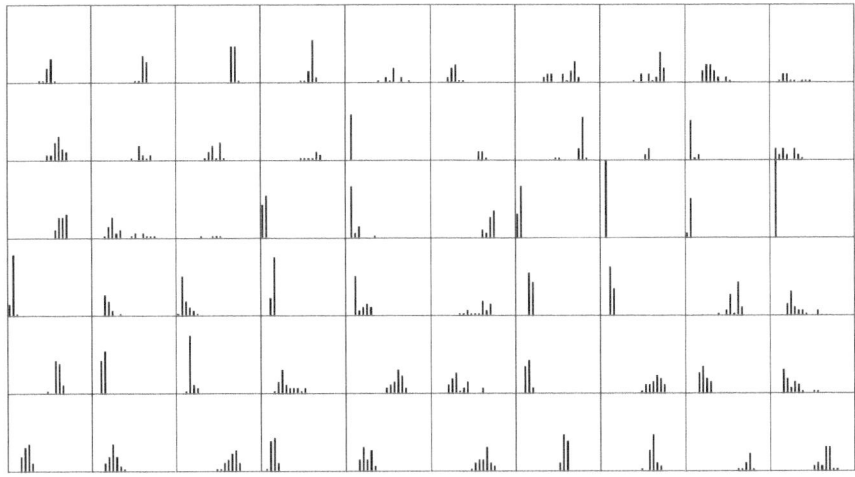

Fig. 1. Histograms on some 5×5 cells.

All six such representative histograms are demonstrated in Figure 3. One may immediately find that the dominant value range for each group is different. This is the fundamental reason in speeding up the searching process to answer a statistical range query.

3.3 A Specific Query Result

Now, let us use the histograms to answer a specific value range query, namely find all high resolution cells on which the NDVI values are between 0.04 and 0.26. Table 1 lists the related values for answering this specific query. First of all, the search is performed on cell groups at the high level (low resolution). The scanning order of groups is determined according to the numbers of high resolution cells in the representative histograms. In fact, the order of representative histograms in Figure 3 is determined based on these numbers. The two solid vertical lines denote the boundaries of the given value range. The number of included cells is high at the top and decreases downwards. The first column in Table 1 is the cluster identifier which is listed in the same order as that in Figure 3. The "cells" column provides the number of low resolution cells we searched in each cluster. The values in the third column present the counts to the query result from each cluster as a percentage. The time saving values and accuracy are also listed in this table.

Time Saving: Time savings result from two effects. One is from the number of empty cells. Since NDVI covers only land, empty cells (cells without valid values) result from regions of water (oceans). For this example, there are a total of $360 \times 180 = 64,800$ low level cells. However, there are only 15,722 non-empty cells. Similarly, there are $72 \times 36 = 2,592$ high level cells, only 887 of which are

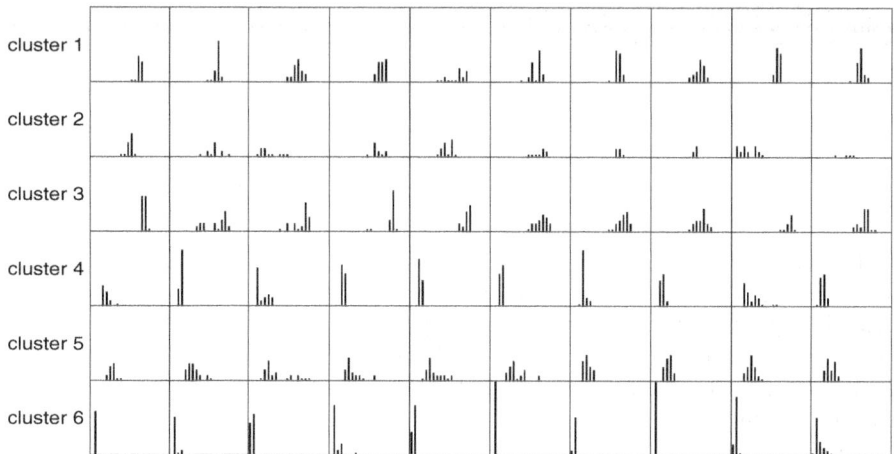

Fig. 2. Histograms grouped by clustering method.

Table 1. Query result. All values except the first two columns are percentage values.

ID	cells count	accuracy	saving 1	saving 2
4	76 40.84	40.84	97.07	91.43
2	458 32.48	73.32	79.4	39.8
5	50 25.46	98.78	77.47	34.16
1	182 0.56	99.34	70.45	13.64
6	26 0.5	99.84	69.44	10.71
3	95 0.16	100	65.78	0

non-empty. When the index system is built, the empty cells are dropped from the problem. Therefore, searching the cells by indices will automatically ignore the empty cells and, correspondingly, save the searching time. One column, labeled saving 1, shown in Table 1 is the percentage time saving rate with empty cells. The last column, saving 2, gives the time saving rate without including the empty cells. It is evident that the time saving from the clustering is significant when only a few groups are counted in the result. The cell numbers in the second column could be used to estimate the time saving. However, as more and more clusters are included, the time saving from clustering drops very quickly. Of course, when all clusters are considered, no time saving at all results from the clustering method. The key point is that only a few clusters would usually give a very high accuracy.

Accuracy: It is obvious that the accuracy increases with the number of clusters included in the search. If we propose a real statistical range query,

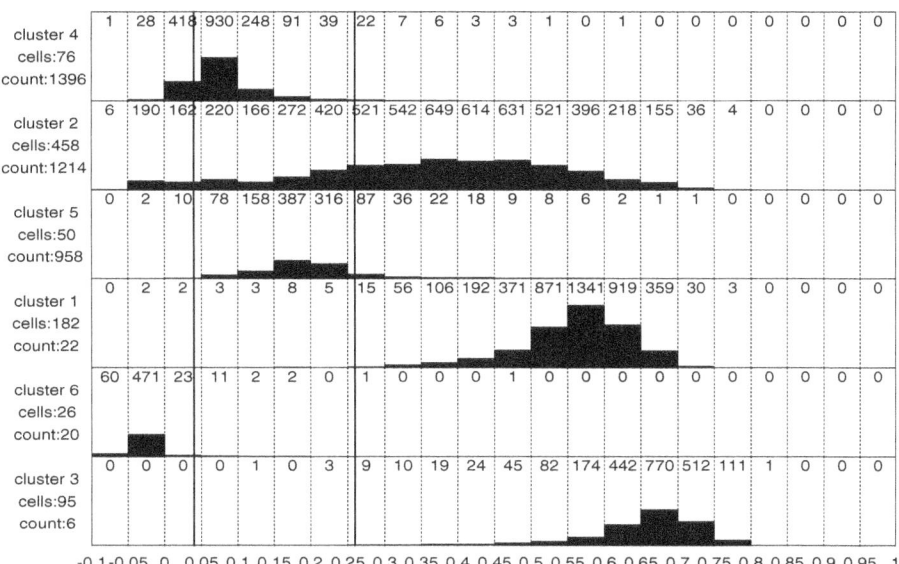

Fig. 3. Representative histograms of six clusters.

say with 95% accuracy, we should reach the desired accuracy after including only three clusters. The contribution from the last three clusters is of very small percentage. As a result, the computing time is saved by working on only the first three groups of cells. The last column in the table indicates that the time saving is significant even without including empty cells. It should be noticed that the time saving from our clustering method is significant for high accuracy but not for full accuracy (100%), when the empty cells are not considered. Nevertheless, the time saving with empty cell consideration still holds even for a query where full accuracy applies.

Effect of number of clusters: As we discussed before, there is no practical algorithm to determine the optimal number of clusters for a problem. To study the influence of cluster number on the query result, we also tried different numbers of clusters for the same problem. Table 2 illustrates the results of this test. The first column is the number of total clusters we choose for the problem. The second column gives the number of selected clusters to obtain at least 95% accuracy. Since the accuracy is not improved continuously, the real accuracy we achieved by including the selected number of clusters should be higher than 95%. These real accuracies are listed in column 3. The last two columns register the time savings with/without including empty cells, respectively, as in Table 1.

It is interesting to note that the results are not very sensitive to the number of clusters that we pick. Moreover, for some number ranges, say $(3,4)$, $(6,7)$ and $(8,9,10)$, the time saving result is the same. The reason is that the change of cluster number does not affect the contents of clusters with values concentrated in

Table 2. Dependence on number of clusters

total	selected	accuracy	saving 1	saving 2
3	2	99.5	66.8	2.93
4	3	99.5	66.8	2.93
5	3	99.3	70.5	13.6
6	3	98.8	77.5	34.2
7	3	98.8	77.5	34.2
8	3	98.6	78.7	37.7
9	3	98.6	78.7	37.7
10	4	98.6	78.7	37.7

the given query value range. Another point is that the time saving, in particular, the time saving without including the empty cells increases with the cluster number. This is not a surprising result because more clusters would result in more accurate distinction of value ranges. Therefore, fewer high level cells need to be scanned for a given value range. Of course, since more clusters would consume more resources in a real system for answering given statistical range queries, we should include only a few clusters for a real problem.

Distance measurement: The effects of different distance measurements on the result are also studied. Table 3 lists results for the Manhattan, Euclidean and maximum distance metrics, which are directly supported by Splus. The results come from the case of NDVI with a 95% given accuracy and a value range between 0.04 and 0.26. The total cells in column two are the total high level cells included to achieve the given accuracy. In all cases, the total number of clusters is 6, and 3 clusters are selected.

The maximum distance method is simply taking the largest difference of all corresponding component differences between two vectors. The maximum metric should not give a very good result since it does not consider the shapes of histograms at all. Nevertheless, it is unexpected that the Euclidean method does not give a result as good as the Manhattan method. This is because the Euclidean metric is more sensitive to outliers. Therefore, most histograms are clustered in one group. This, of course, reduce the efficency of the clustering method. Since the shape of histograms is the most important criterion for clustering, Q-type measurements[20] such as correlations would possibly lead to a better result. More studies are needed to confirm this.

Intuitive result map: To intuitively demonstrate the accuracy of the query result, a map with the selected cells is displayed in Figure 4. The boxes in this figure are non-empty 5 × 5 cells. The cells denoted by heavy solid lines are those selected for the range query, and those with thin line boundaries are ignored. The dots are 1 × 1 cells on which the parameter values within the defined range fall. Most of the dots are in thick boxes as expected. But, there exist some 1 × 1 cells satisfying the query condition but ignored in the result such as points near the upper-left corner. Certainly, there are some selected high level cells on which no

Table 3. Dependence on distance measurement methods

method	total cells	accuracy	saving 1	saving 2
Manhattan	584	98.8	77.5	34.2
Euclidean	855	97.5	67.0	3.61
maximum	848	98.5	67.3	4.40

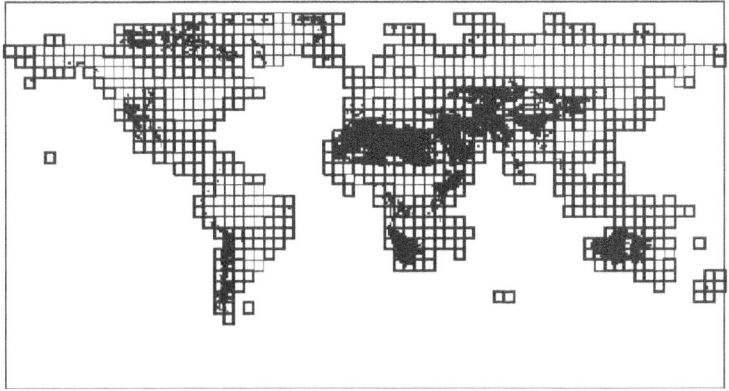

Fig. 4. Map of selected cells and ignored cells.

low level cell is found to be in the selected range. That is because the clustering is based on distances between histograms and all cells in a cluster instead of individual cells are marked as selected once the cluster is covered.

3.4   Another Query

Table 4 is similar to Table 1, but gives the result for another value range, namely [0.56, 0.78]. Since the searching order is computed according to the value range, this order is of course different from the order in Table 1. However, the efficiency is almost the same. After including three clusters of cells in the search, the accuracy is already more than 99%. If the given accuracy is 95%, the time saving should be 71% and 17% for the situation with and without including empty cells, respectively. In fact, the searching order could be estimated from Figure 3, the representative histograms. Figure 3 shows the histograms in descending order of counts of high resolution cells for the range query between 0.04 and 0.26. The results are absolutely the same as that given in Table 1. For the value range between 0.56 and 0.78, one could guess the order according to this figure to be cluster 1, cluster 3, and cluster 2, the same as those shown in Table 4.

Table 4. Query result for NDVI values between 0.56 and 0.78.

ID	cells count	accuracy	saving 1	saving 2
1	182 47.35	47.35	92.98	79.48
3	95 37.99	85.35	89.31	68.77
2	458 14.44	99.79	71.64	17.14
5	50 0.19	99.98	69.71	11.5
4	76 0.02	100	66.78	2.93
6	26 0	100	65.78	0

3.5 Data without Empty Cells

We discussed before that the time saving for answering statistical range queries arise from the effects of both ignoring empty cells and the clustering method itself. To constrain the contribution of empty cells, we use this method to a data set without empty cells (except for a few missing values). The data set is a 1×1 global monthly mean surface skin temperature data[18]. The data format and structure are the same as those of NDVI data we used. Table 5 shows an example result for a query range between 250 and 280 Kelvin. Here, time saving is fully from the clustering method itself, and the amount is very significant. As with the parameter NDVI, we can also study the effect of the cluster number. The general trend is the same. However, in this case, the last 4 cases with cluster number from 7 to 10 give the same result for 95% accuracy. The time saving is 63.5%.

Table 5. Result for data without empty cells.

ID	cells count	accuracy	saving 1	saving 2
4	481 36.86	36.86	81.44	81.44
5	290 35.93	72.79	70.25	70.25
1	541 26.94	99.73	49.38	49.38
2	318 0.23	99.95	37.11	37.11
3	670 0.05	100	11.27	11.27
6	292 0	100	0	0

4   Discussion and Future Work

The clustering method described here is used to answer statistical range queries in Earth system science quickly and with a satisfactory accuracy assessment. The time saving rates depend on many factors such as data type, data resolution, number of clusters, selected value range, and even data value distribution.

Gridded scientific data sets are usually of uniform internal structure with missing values filled by non-valid values. Therefore, using the time saving rates with empty-cell consideration is not an unrealistic estimate. The results shown above indicates that the time saving factor for examples with the 1×1 degree resolution NDVI data is around 4. That is, the searching time needed with clustering method is only about one fourth of the searching time without using this method. When one considers the NDVI data sets with $1km \times 1km$ resolution (much higher resolution), the time saving factor of the clustering method could be more than a factor of ten since the cell compactness at the high level will be high. If time dimension is added to the problem, the time saving factor will be even higher.

To handle very high resolution data sets, it is better to have similar numbers of data values in different cells. This requires that each cell covers regions with valid data values only. For example, if data values cover land only, we should not include ocean parts in cells in coastal regions. The clustering method works for this case. Although we use gridded data sets in latitude-longitude coordinate system, the cell division is not necessary to follow the lat-lon lines. The only requirement for dividing cells is that each high resolution (low level) values should associate with one and only one high level cell. In other words, the cell may have arbitrary shapes as long as no overlapping takes place. Coastal lines may be one natural boundary definition for cells. Another natural choice is political boundaries. Of course, other tools, such as GIS, are needed to achieve these types of cell divisions.

To include temporal dimension in the clustering method, the representative histograms should be built on-the-fly. The indexing part should be the same. However, as a query is issued for a certain time period, the representative histograms should be created by summing up individual histograms in each cluster for that time period only. The same procedure should be used for queries in selected spatial areas instead of the whole covered area. R-tree indices could also be used for this purpose[21].

The clustering method works because the representative histograms are of different dominant ranges. In other words, for a given value range, there may exist a large amount of associated cells on which the parameter values are in the given range, However, most of such associated cells fall in just a few clusters. Other clusters may contain a few associated cells or none at all. This feature makes the clustering method most efficient for data of continuously distributed values. Most scientific data sets are of this property but values in commercial databases may lack it. Therefore, users could get the most benefit by using this method on scientific data sets, especially remote sensing data for the Earth.

A statistical range query result may be more useful in actual applications than a full accuracy result for scientific interests. If we give a relatively high accuracy requirement, say 95%, most interesting regions should be selected by a statistical range query. Values in the given range but not selected are usually isolated or sparse in the distribution. Actually, it is desired not to include isolated points or tiny regions in the query result. Ideally, a user should give an estimate size of area as part of query restriction, only areas larger than the given size should be returned to the user. That could be one direction we should work on in the future. Another improvement is to use continuous density functions instead of histograms. Korn et al.[22] employed this technique for range queries on continuous scientific data. The same technique can be used for the clustering method as well.

For a database with incremental data sets, using predefined fixed minimum and maximum values to define a fixed bin frame may cause problems when new data sets have values outside the predefined range. Reasonable value ranges are well defined for scientific data sets. To deal with the new extreme values, we can define the bins on the two ends with infinite sizes. That is, we define the first bin covers values less than a given value (a reasonable minimum data value plus an increment of regular bin size) and the last bin spreads all values larger than another predefined value (a reasonable maximum data value less an increment of regular bin size). Though an error may result in the distance computations due to this definition, the amount should be very limited.

One interesting use of this method is to develop a data information system supporting the statistical range queries. Some dynamical data query and analysis systems have been developed[23, 24, 25]. The clustering method and interactive control of query process could make the system very useful. An online data exploration system was presented during the 11th Scientific and Statistical Database Management System conference[6]. In this example, the accuracy is improved by including more sample entries from a database as time progresses, and users can interrupt a search at will as the query result achieves some sufficient accuracy. In the clustering method, the aggregation values (histograms) are precomputed and used for indexing. When a range query is issued, the result could be returned very quickly but with low accuracy. As more and more clusters are included, higher and higher accuracy will be obtained. If the intermediate result is piped back to a user interface and the system allows users to control the query process, we should have similar appearance effect as that for the on-line aggregation method[6] though the mechanisms for the two systems are different underneath. Undoubtedly, in this case, accuracy increments are not continuous.

References

[1] G. Asrar and R. Greenstone, eds., *1999 EOS Reference Handbook.* NASA (Washington, DC), 1999.

[2] NASA's ESIP, "Earth Science Information Partners," 2000.
http://www.esipfed.org/.

[3] Content-based Search and Data Mining Cluster of ESIP, "Science Scenarios for Content-based Search and Data Mining," 2000.
http://esipfed.org:8080/Clusters/Content_Based/sci_scen.html.

[4] J. Gary, A. Bosworth, A. Layman, and H. Pirahesh, "Data Cube: A Relational Aggregation Operator Generalizing Group-by, Cross-tabs, and Sub-totals," in *Proceedings of IEEE Conf. on Data Engineering*, pp. 152–159, 1996.

[5] J. M. Hellerstein, P. J. Haas, and H. J. Wang, "Online Aggregation," in *Proce. 1997 ACM SIGMOD Intl. Conf. Management of Data*, pp. 171–182, ACM Press, 1997.

[6] P. J. Haas, "Techniques for Online Exploration of Large Object-Relational Databases," in *Proceedings of the 11th International Conference on Scientific and Statistical Database Management* (Z. M. Ozsoyoglu, G. Ozsoyoglu, and W.-C. Hou, eds.), pp. 4–12, IEEE, Computer Society, 1999.

[7] V. Poosala, Y. Ioannidis, P. Haas, and E. Shekita, "Improved Histograms for Selectivity Estimation of Range Predicates," in *Proce. 1996 ACM SIGMOD Intl. Conf. Management of Data*, pp. 294–305, ACM Press, 1996.

[8] V. Poosala and V. Ganti, "Fast Approximation Answers to Aggregate Queries on a Data Cube," in *Proceedings of the 11th International Conference on Scientific and Statistical Database Management* (Z. M. Ozsoyoglu, G. Ozsoyoglu, and W.-C. Hou, eds.), pp. 24–33, IEEE, Computer Society, 1999.

[9] D. Carr and A. R. Olsen, "Simplifying Visual Applearance by Sorting: An Example Using 159 AVHRR Classes," *Statistical Computing & Statistical Graphics Newsletter*, pp. 10–16, April 1996.

[10] A. K. Jain, M. N. Murty, and P. J. Flynn, "Data Clustering: A Review," *ACM Computing Surveys*, vol. 31, no. 3, pp. 264–323, 1999.

[11] B. Everitt, *Cluster Analysis*. John Wiley & Sons, 1993.

[12] J. Puzicha, T. Hofmann, and J. M. Buhmann, "Histogram Clustering for Unsupervised Segmentation and Image Retrieval," *Pattern Recognition Letters*, vol. 20, pp. 899–909, 1999.

[13] Z. Li, X. S. Wang, M. Kafatos, and R. Yang, "A Pyramid Data Model for Supporting Content-based Browsing and Knowledge Discovery," in *Proceedings of the 10th International Conference on Scientific and Statistical Database Management* (M. Rafanelli and M. Jarke, eds.), pp. 170–179, IEEE, Computer Society, 1998.

[14] W. Venables and B. Ripley, *Modern Applied Statistics with S-Plus*. Springer-Verlag, 1994.

[15] R. T. Ng and J. Han, "Efficient and Effective Clustering Methods for Spatial Data Mining," in *Proce. of the 20th VLDB Conference Santiago, Chile*, pp. 144–155, 1994.

[16] T. Zhang, R. Ramakrishnan, and M. Livny, "BIRCH: An Efficient Data Clustering Method for Very Large Databases," *SIGMOD Rec.*, vol. 25, no. 2, pp. 103–114, 1996.

[17] Z. Huang, "Extension to the k-Means Algorithm for Clustering Large Data Sets with Categorical Values," *Data Mining and Knowledge Discovery*, vol. 2, no. 3, pp. 283–304, 1998.

[18] H. Kyle, J. McManus, S. Ahmad, and et al., *Climatology Interdisciplinary Data Collection, Volumes 1-4, Monthly Means for Climate Studies*. NASA Goddard DAAC Science Series, Earth Science Enterprise, National Aeronautics & Space Administration, NP-1998(06)-029-GSFC, 1998.

[19] A. P. Cracknell, *The Advanced Very High Resolution Radiometer*. Taylor & Francis Inc., 1997.

[20] J. D. Jobson, *Applied Multivariate Data Analysis*, vol. 2. Springer, 1992.

[21] A. Guttman, "R-trees: A Dynamic Index Structure for Spatial Searching," in *Proc. ACM SIGMOD*, pp. 47–57, June 1984.

[22] F. Korn, T. Johnson, and H. V. Jagadish, "Range Selectivity Estimation for Continuous Attributes," in *Proceedings of the 11th International Conference on Scientific and Statistical Database Management* (Z. M. Ozsoyoglu, G. Ozsoyoglu, and W.-C. Hou, eds.), pp. 244–253, IEEE, Computer Society, 1999.

[23] M. Kafatos, Z. Li, R. Yang, and et al., "The Virtual Domain Application Data Center: Serving Interdisciplinary Earth Scientists," in *Proceedings of the*

9th International Conference on Scientific and Statistical Database Management (D. Hansen and Y. Ioannidis, eds.), pp. 264–276, IEEE, Computer Society, 1997.

[24] M. Kafatos, X. Wang, Z. Li, R. Yang, and D. Ziskin, "Information Technology Implementation for a Distributed Data System Serving Earth Scientists: Seasonal to Interannual ESIP," in *Proceedings of the 10th International Conference on Scientific and Statistical Database Management* (M. Rafanelli and M. Jarke, eds.), pp. 210–215, IEEE, Computer Society, 1998.

[25] R. Yang, C. Wang, M. Kafatos, X. Wang, and T. El-Ghazawi, "Remote Data Access via SIESIP Distributed Information System," in *Proceedings of the 11th International Conference on Scientific and Statistical Database Management* (Z. M. Ozsoyoglu, G. Ozsoyoglu, and W.-C. Hou, eds.), p. 284, IEEE, Computer Society, 1999.

Acknowledgment

We acknowledge partial prototype funding support from the NASA ESDIS Project (NAG5-8607), from the Earth Science Enterprise WP-ESIP CAN Program as well as from George Mason University.

Fast Randomized Algorithms for Robust Estimation of Location

Vladimir Estivill-Castro[1] and Michael E. Houle[2]

[1] Department of Computer Science & Software Engineering,
The University of Newcastle, Callaghan, NSW 2308, Australia.
vlad@cs.newcastle.edu.au
[2] Basser Department of Computer Science
The University of Sydney, Sydney, NSW 2006, Australia.
meh@cs.usyd.edu.au

Abstract. A fundamental procedure appearing within such clustering methods as k-Means, Expectation Maximization, Fuzzy-C-Means and Minimum Message Length is that of computing estimators of location. Most estimators of location exhibiting useful robustness properties require at least quadratic time to compute, far too slow for large data mining applications. In this paper, we propose $O(Dn\sqrt{n})$-time randomized algorithms for computing robust estimators of location, where n is the size of the data set, and D is the dimension.

Keywords: clustering, spatial data mining, robust statistics, location.

1 Introduction

When analyzing large sets of spatial information (both 2-dimensional and higher-dimensional), classical multivariate statistical procedures such as variable standardization, multivariate studentizing, outlier detection, discriminant analysis, principal components, factor analysis, structural models and canonical correlations all require that the center and scatter of a cloud of points be estimated [17, 19].

Estimating the center of a cloud of points is also the fundamental step of iterative algorithms for spatial clustering under several inductive principles. An important example is that of maximum likelihood estimation, which the Expectation Maximization (EM) heuristic uses to compute parameters of location and scatter. A close relative is the well-known k-Means iterative procedure, where cluster centers are computed as the arithmetic mean of their elements. The quality of these iterative approaches (and others such as Minimum Message Length [18], Fuzzy-C-Means [2], and Vector Quantization [2]) is directly dependent upon the quality of the methods for estimating location and scatter. It is worth noting that robust estimation of location is typically a prerequisite for robust estimation of scatter.

In practice, a very common way of estimating location and scatter of groups of observations is by using their means and variances, under the assumption

J.F. Roddick and K. Hornsby (Eds.): TSDM 2000, LNAI 2007, pp. 77–88, 2001.

that the observations have underlying normal distributions. Also, when comparing independent groups, it is often assumed that the distributions have a common variance. However, hundreds of journal articles in statistics have described serious practical difficulties in those situations in which the groups differ; in particular, the sample mean can have a relatively large standard error under slight departures from normality. For the geo-referenced data typically maintained in current geographical information systems (GIS), the assumption of normality is almost always unrealistic.

Since the 1960s, much attention has been given to the robustness of statistics used in the analysis of variance and regression [19]. Intuitively, estimation of location and scatter is robust when the statistic computed from the data is affected only very slightly under arbitrary but small shifts from the assumptions on the underlying distribution, or by the presence of outliers. Generally speaking, the use of robust statistics allows for a departure from the assumption of normality, a reduction in the standard error in location statistics, tighter confidence intervals, and more success in identifying differences when standard hypothesis testing fails.

Of the notions of robustness that appear in the statistical literature, we will be interested in two in particular — infinitesimal robustness and quantitative robustness — both having precise mathematical formulations [7, 8]. A number of such robustness properties have been proven for affine equivariant estimators of location (measures invariant under translation and scaling [17]):

- good local sensitivity properties (as measured by an appropriate influence function),
- good global sensitivity properties,
- $n^{1/2}$-consistency, where n is the number of data points,
- asymptotically normal distribution,
- statistical efficiency, for a broad class of underlying model distributions that include the multivariate normal.

Many estimators have been proposed as being robust, in that they exhibit some of the robustness properties listed above; however, none have been shown to be superior in all aspects. Much effort has gone into showing whether a particular estimator exhibits a particular robustness property [17], but relatively little attention has been given to the analysis of the feasibility or complexity of computing these estimators. For the large volumes of data that arise in data mining applications, it is crucial that only computationally efficient robust estimators of location be used.

In this paper, we propose a new family of non-deterministic (yet robust) estimators of location that rely on fast randomization techniques for their computation. The non-deterministic estimators approximate in subquadratic time certain robust deterministic estimators which require quadratic time to compute exactly. The efficiency of the non-deterministic estimation makes it more amenable in spatial data mining settings involving multivariate data.

2   Orthogonal Equivariant Estimators

Statisticians believe that outliers are much harder to identify in multivariate data clouds than in the univariate case [14]. To this end, a large family of robust multivariate estimators of location have been proposed [14]. Statisticians have shown particular interest in estimators of location that are affine equivariant. This means that if $X = \{x_1, ..., x_n\} \subset \Re^D$ is a set of points regarded as a cloud in D-dimensional real space, the estimator T that finds the 'center' of the cloud satisfies

$$T(Ax_1 + b, \ldots Ax_n + b) = AT(x_1, \ldots x_n) + b,$$

for all non-singular matrices A and vectors $b \in \Re^D$. That is, T is invariant with respect to linear transformation.

In this paper we concentrate on orthogonal equivariant estimators — that is, those that are equivariant to linear transformations that preserve Euclidean distances (reflections and rotations as well as translations), and correspond to the case where the matrix A is orthogonal (that is, $A^T = A^{-1}$).

We concentrate on this family of estimators for two reasons. First, the computational complexity of orthogonal estimators changes significantly when we progress from the one-dimensional setting to the multi-dimensional setting. Second, equivariant orthogonal estimators seem to be a natural stepping-stone towards the development of algorithms for the more general class of multi-dimensional affine estimators. There are three families of multi-dimensional affine estimators [17] to consider: M-estimators (both monotone and non-monotone), high breakdown-point estimators, and the family of S-estimators. Previously existing algorithms for all three families are too inefficient for the data sets that arise in Knowledge Discovery and Data Mining (KDDM) applications.

Our work is based on two estimators with maximum (optimal) break-point robustness: the Least Median Square (LMS) and Least Trimmed Square (LTS) estimators [14]. In order to present these robust estimators of location, we must first introduce two operators. For a set $S \in \Re$ of s numbers, let MED(S) denote the median of S, which we shall define as the $(\lfloor \frac{n}{2} \rfloor + 1)$st value in the sorted order of the elements of S. The second operator is the radius of the smallest ball centered at x that includes at least half of the points in X, and is defined by HALF_BALL_RAD(x) = MED$(\{d(x, x_i)|x_i \in X\})$, where d is the Euclidean distance $d(x, y) = \sqrt{(x - y) \cdot (x - y)}$.

The orthogonal estimators of location LMS(X) and LTS(X) are defined as follows:

- A point x determines LMS(X) if and only if x minimizes

$$\text{LMS}(x) = \text{HALF\_BALL\_RAD}(x)^2.$$

- A point x determines LTS(X) if and only if x minimizes

$$\text{LTS}(x) = \sum_{d(x, x_i) \leq \text{HALF\_BALL\_RAD}(x)} d(x, x_i)^2.$$

Subquadratic algorithms for these two robust estimators of location are known only for the unidimensional case ($D = 1$).

Theorem 1. *[14] Let $\langle x_1, \ldots x_n \rangle$ denote the unidimensional data in sorted order. Let I_j denote the interval $[x_j, x_{\lfloor n/2 \rfloor + j}]$, for $j = 1, \ldots, \lceil n/2 \rceil$. Let $l(I_j) = x_{\lfloor n/2 \rfloor + j} - x_j$ be the length of the interval I_j. Let $I_{j_{\min}}$ be the interval with smallest length. Then, $\mathrm{LMS}(X) = (x_{j_{min}} + x_{\lfloor n/2 \rfloor + j_{\min}})/2$.*

Note that the theorem is slightly different if $\mathrm{MED}(S)$ is defined as the midpoint of the $n/2$-th item and the $n/2 + 1$-th item, in the case when n is even.

In this one-dimensional setting, sorting X and then scanning the list for the minimum-length interval $I_{j_{\min}}$ results in an $O(n \log n)$ algorithm for determining $\mathrm{LMS}(X)$. Similarly, $\mathrm{LTS}(X)$ can be computed in $O(n \log n)$ time by sorting X, and then computing for all $j = 1, \ldots, \lfloor n/2 \rfloor$ the following:

- the average $\mathrm{AVG}(X \cap I_j) = \sum_{k=j}^{\lfloor n/2 \rfloor + j} x_k / (\lfloor n/2 \rfloor + 1)$ of those data values in I_j, and
- the sum of squared differences $\mathrm{SQS}(X \cap I_j) = \sum_{k=j}^{\lfloor n/2 \rfloor + j} d(x_k, \mathrm{AVG}(X \cap I_j))^2$ between this average and the data values in I_j.

$\mathrm{LTS}(X)$ is simply that average value $\mathrm{AVG}(X \cap I_j)$ associated with the smallest sum of squares $\mathrm{SQS}(X \cap I_j)$. Note that $\mathrm{SQS}(X \cap I_{j+1})$ can be computed from $\mathrm{SQS}(X \cap I_j)$ in constant time, and thus the computation of $\mathrm{LTS}(X)$ requires only linear time after the initial $O(n \log n)$ sorting step.

3   Randomized Algorithms for Estimation

In the multidimensional case ($D > 1$), there seems to be no subquadratic algorithm for the continuous optimization that LMS and LTS demand. One natural alternative to consider then is a discrete variant of the problem. For the remainder of this paper, we will restrict our attention to the discrete estimators $\mathrm{LMS\_D}(X)$ and $\mathrm{LTS\_D}(X)$ minimizing $\mathrm{LMS}(\boldsymbol{x})$ and $\mathrm{LTS}(\boldsymbol{x})$ over all choices of $\boldsymbol{x} \in X$ only.

$\mathrm{LMS\_D}(X)$ and $\mathrm{LTS\_D}(X)$ can be computed in $O(Dn^2)$ time by simply evaluating for each $\boldsymbol{x}_i \in X$ either $\mathrm{LMS}(\boldsymbol{x})$ or $\mathrm{LTS}(\boldsymbol{x})$, as appropriate, returning the $\boldsymbol{x}_i$ that achieves the minimum. The calculation of $\mathrm{LMS}(\boldsymbol{x})$ or $\mathrm{LTS}(\boldsymbol{x})$ requires $O(Dn)$ time, as detailed in the following steps:

1. Compute $d(\boldsymbol{x}, \boldsymbol{x}_i)^2$ for $i = 1, \ldots, n$, in $O(Dn)$ time.
2. Compute the median of the n values found in the previous step, in $O(n)$ time.

When computing $\mathrm{LTS}(\boldsymbol{x})$, the values smaller than the median can be totaled in $O(n)$ time.

Since algorithms with quadratic time complexity are unacceptable for most KDDM applications [6, 10], we propose two randomized algorithms that seek to approximate the estimators $\mathrm{LMS\_D}(X)$ and $\mathrm{LTS\_D}(X)$ in subquadratic expected time. For more information about the theory of randomized algorithms, see [11].

3.1   Random Sampling Method

With respect to a random sample R of r points, given some $x \in X$, we define the non-deterministic estimator LMS_S(x) to be the value one would obtain if LMS_D(x) were computed after first replacing every point in X by its nearest neighbour in R. Similarly, we may define the estimator LTS_S(x).

The algorithm which follows, RS, computes a non-deterministic estimator LMS_S(X) approximating LMS_D(X), by first generating a random sample $R \subset X$ and computing estimators LMS_S(x) for all $x \in X$. Those points of X yielding the u smallest values of LMS_S(x) are then selected for a second round of computation, in which LMS_D(x) is computed exactly over the full set X. LMS_S(X) is taken to be the smallest value of LMS_D(x) obtained in this second round. The same method can be used to compute a non-deterministic estimator LTS_S(X) approximating LTS_D(X).

It should be noted that both LMS_S(X) and LTS_S(X) are D-dimensional random variables. However, LMS_S(X) = LMS_D(X) whenever the point determining LMS_D(X) happens to generate one of the u best values of LMS_S(x). The same holds true for LTS_S(X) and LTS_D(X).

Algorithm RS

1. Select a subset $R = \{y_1, \ldots, y_r\}$ uniformly at random from among all subsets of X of size r. This requires $O(r)$ time [13].
2. For each $x \in X$, find its nearest element in R. Let $C_i = \{x \in X \mid y_i$ is the nearest element of R to $x\}$. The sets $C_1, \ldots, C_r$ constitute a partition of X. Note that $|C_i|$, $\sum_{x \in C_i} d(x, y_i)$, $\min_{x \in C_i} d(x, y_i)$ and $\max_{x \in C_i} d(x, y_i)$ can all computed incrementally in $O(Drn)$ total time without complex data structures.
3. With the information of the previous step it is possible to compute LMS_S(x) or LTS_S(x) in $O(r)$ time.
   - Compute $\delta_i = d(x, y_i)$ for $i = 1, \ldots, r$ in $O(r)$ time.
   - Find the median δ_m of the values δ_i, each with weight $|C_i|$. This can be done in $O(r)$ time. Set LMS_S$(x) = \delta_m$.
   - If we wish to compute LTS_S(X), then let $I = \{i| \delta_i < \delta_m\}$ be the index set of points with distances less than the median. Set

$$\text{LTS\_S}(x) = \sum_{i \in I}^{k} |C_i| \cdot \delta_i + \delta_m \left(\left\lfloor \frac{n}{2} \right\rfloor + 1 - \sum_{i \in I}^{k} |C_i| \right).$$

   The summations can be computed in $O(r)$ time.
4. Compute LMS_S(x) or LTS_S(x) for all $x_i \in X$, as appropriate. This requires $O(rn)$ total time.
5. Find the smallest u approximation values computed in the previous step. Denote by U the set of points of X corresponding to these u values. The time cost for this step is $O(un)$.
6. For all $x \in U$, evaluate LMS_D(x) or LTS_D(x) exactly. Return the point in U corresponding to the smallest value. This requires $O(Dun)$ total time.

As the total cost of Algorithm RS is $O(Dn \cdot \max\{u, r\})$ time, it is sensible to choose $r = \Theta(u)$. The issues surrounding the choice of u will be discussed later.

3.2   Random Partition Method

In contrast to the sampling method, which uses approximations of LMS_D(x) to generate a small list of candidates for which the exact method is applied, our second algorithm RP generates its small list of candidates by first randomly partitioning X into smaller subsets $Y_1, Y_2, \ldots, Y_u$, and then adding to the list the point which determines the value of LMS_D(Y_i) with respect to Y_i. Each point μ generated is then fed into a second round of computation as before, in which LMS_D(μ) is computed exactly with respect to X, and the smallest value chosen. Algorithm RP can be used to produce an estimator LTS_D(X) in much the same way.

Let LMS_P(X) and LTS_P(X) denote the non-deterministic estimators computed by the RP method. LMS_P(X) = LMS_D(X) whenever the point determining LMS_D(X) happens to determine the value with respect to the subset to which it is assigned by the algorithm. The same holds true for LMS_P(X) and LMS_D(X) .

Algorithm RP

1. Generate a permutation of X uniformly at random; let the resulting sequence be $Y = \langle y_1, \ldots, y_n \rangle$. This can be performed in $O(n)$ time [13].
2. Let $u \in \{1, \ldots, n\}$ be an integer parameter. We consider a partition of the sequence Y into u consecutive blocks $Y_1, Y_2, \ldots, Y_u$, each containing roughly $\frac{n}{u}$ elements.
3. For all $1 \leq i \leq u$, compute the values LMS_D(Y_i) or LTS_D(Y_i) with respect to Y_i, using the exhaustive quadratic time method described at the beginning of this section. Each of the r computations requires $O(D\frac{n^2}{u^2})$ time, for a total of $O(D\frac{n^2}{u})$. Let $\mu_i \in Y_i$ be the point determining the value for Y_i.
4. For all $1 \leq i \leq u$, compute the values LMS_D(μ_i) or LTS_D(μ_i) with respect to X, and return the μ_i corresponding to the smallest value computed. This requires $O(Dun)$ total time.

The total cost of Algorithm RP is $O(Dn \cdot \max\{u, \frac{n}{u}\})$ time. This is minimized when $u = \Theta(\sqrt{n})$.

4   Statistical Robustness

4.1   Influence Curve

The first measure of robustness we consider derives from the *influence curve* [19, page 6], which describes the effect of adding a new element to an existing sample. As an example of an influence curve, consider how the arithmetic mean of a set of n real values $\{x_1, \ldots, x_n\}$ is modified by the addition of a new point x_{n+1}:

$$\text{AVG}(\{x_1, \ldots, x_n, x_{n+1}\}) = \frac{n}{n+1}\text{AVG}(\{x_1, \ldots, x_n\}) + \frac{1}{n+1}x_{n+1}.$$

Thus, if we take n and the points $x_1, x_2, \ldots, x_n$ to be fixed quantities, and the point x_{n+1} as a variable t, the arithmetic mean of the extended sample becomes the linear function

$$\text{AVG}(t) = c_1 + c_2 t$$

for suitable constants c_1 and c_2. The unboundedness of $\text{AVG}(t)$ indicates that there is no limit on the effect that a single outlier can have on the mean. On the other hand, one would expect any robust statistic (such as the arithmetic median) to have an influence curve which is bounded.

Our two algorithms are randomized, and as such can produce a different result each time they are executed. However, despite their randomization, the estimators produced are provably robust. For both algorithms, we will assume that the parameters are chosen to be $r = u = \lceil \sqrt{n} \rceil$.

Proposition 1. *Let X be a sample of n points drawn from $\Re^D$, for $n \geq 2$. Then the influence curves of* LMS_S(X), LMS_P(X), LTS_S(X) *and* LTS_P(X) *are all bounded.*

Proof. The proof is only shown for the case of LMS_P(X); for the other cases the arguments are similar.

Consider the behavior of RP when a new point $\boldsymbol{x}'$ is introduced into X. If LMS_P$(X \cup \{\boldsymbol{x}'\}) \in X$, then the result holds, as X is assumed to be fixed. Let us therefore assume that LMS_P$(X \cup \{\boldsymbol{x}'\}) = \boldsymbol{x}'$.

For this to happen, RP must have computed LMS_D$(\boldsymbol{x})$ exactly for $\lceil \sqrt{n+1} \rceil$ different choices of $\boldsymbol{x} \in X \cup \{\boldsymbol{x}'\}$, including $\boldsymbol{x} = \boldsymbol{x}'$. Let $\boldsymbol{z} \neq \boldsymbol{x}'$ be one of these choices, and let δ_z and $\delta_{x'}$ be the values of LMS_D$(\boldsymbol{z})$ and LMS_D$(\boldsymbol{x}')$ with respect to $X \cup \{\boldsymbol{x}'\}$, respectively. Let δ be the diameter of X.

If $\delta_{x'} > \delta_z$, then RP would have chosen $\boldsymbol{z}$ (or some other point of X) instead of $\boldsymbol{x}'$. Therefore $\delta_{x'} \leq \delta_z$. In the computation of LMS_D$(\boldsymbol{z})$ with respect to $X \cup \{\boldsymbol{x}'\}$, the closest $\lfloor \frac{n+1}{2} \rfloor + 1$ distances are considered; for $n \geq 2$ there are at least 2 and no more than n of these. Since at least n of the distances must be no more than the diameter of X, we have $\delta_{x'} \leq \delta_z \leq \delta$. Therefore $\boldsymbol{x}'$ must be within distance δ of some point of X, and thus cannot be unbounded. □

4.2 Breakpoint

Returning to our example of the mean and median of a unidimensional set X of n elements, modifying a single data point could result in an arbitrarily-large change in the mean. However, for the median to change arbitrarily, at least $\lceil \frac{n}{2} \rceil$ points would need to be replaced. The *breakpoint* of an estimator is defined as the infimum of the proportion of data elements which can be changed without resulting in an arbitrarily-large change in the estimator, taken over all finite data sets X drawn from the underlying distribution. Thus, the unidimensional mean has a breakpoint of 0, and the unidimensional median has a breakpoint of $\frac{1}{2}$. In general dimensions, it is known that no estimator of location can have breakpoint greater than $\frac{1}{2}$ [14]. Intuitively, when an estimator of location has breakpoint p, then the estimator would still reveal location even if $\lfloor pn \rfloor$ data elements were outliers.

The most attractive property of LMS and LTS is that they are robust with highest possible breakpoint, $\frac{1}{2}$. The same turns out to be true for LMS_P(X) and LTS_P(X).

Proposition 2. LMS_P(X) *and* LTS_P(X) *are robust with breakpoint* $\frac{1}{2}$.

Proof. Again, we show the proof only for LMS_P(X), as the proof for LTS_P(X) is analogous. It suffices to prove that regardless of the random choices made by RP, over all choices of X' such that $|X| = |X'|$ and $|X' \setminus X| = \lceil \frac{n}{2} \rceil - 1$, LMS_P$(X')$ must be bounded when X is fixed.

Let $Y_1, Y_2, \ldots, Y_u$ be the u blocks into which X' is partitioned by RP. Since $|X' \setminus X| = \lceil \frac{n}{2} \rceil - 1 < \frac{n}{2}$, there must be at least one block (call it Y_i) which has more points of X than it does of $X' \setminus X$.

Let n_i be the number of points of Y_i. The number of points of $X \cap Y_i$ must therefore be at least $\lfloor \frac{n_i}{2} \rfloor + 1$. This implies that for each $\boldsymbol{x} \in X \cap Y_i$, the value LMS_D$(\boldsymbol{x})$ with respect to Y_i must be at most δ_i, the diameter of $X \cap Y_i$. The fact that the majority of points of Y_i are in $X \cap Y_i$ thus forces the point $\boldsymbol{y}$ determining LMS_D(Y_i) to be within distance δ_i of some point of $X \cap Y_i$.

If δ is the diameter of the original set X, then $\boldsymbol{y}$ must be within distance $\delta + \delta_i$ of every point of X. In the final step of RP, LMS_D$(\boldsymbol{y})$ is calculated with respect to the full set X'. Since $\lfloor \frac{n}{2} \rfloor + 1$ points of X' are also in X, the value calculated must be at most $\delta + \delta_i$. This implies that LMS_P(X') has distance at most $\delta + \delta_i \leq 2\delta$ to at least $\lfloor \frac{n}{2} \rfloor + 1$ points of X', of which at least one must also be a point of X. Therefore LMS_P(X') is bounded when X is fixed. $\qquad\square$

Unfortunately, the estimators LMS_S(X) and LTS_S(X) turn out to have a breakpoint of 0 whenever the candidate list size u is such that $\lim_{n \to \infty} \frac{u}{n} = 0$. If $|X' \setminus X| \geq u$, it is possible that all candidates on list U could be points of $X' \setminus X$, and thus LMS_S(X') and LTS_S(X') could be arbitrarily far from LMS_S(X) and LTS_S(X).

Nevertheless, even if $|X' \setminus X| = \lceil \frac{n}{2} \rceil - 1$, the probability of this happening is less than $\frac{1}{2^u}$, which diminishes very rapidly with u regardless of the underlying distribution of X. If even one point of U should happen to fall in X, arguments similar to the proof of Proposition 2 show the boundedness of LMS_S(X') and LTS_S(X'). For such choices as $u = \lfloor \sqrt{n} \rfloor$ and $n = 400$, the chance of failure would be less than one in one million: the performance of LMS_S(X) and LTS_S(X) on larger data sets would be effectively that of a robust estimator with breakpoint $\frac{1}{2}$. In this sense, the estimators can be viewed as 'asymptotically robust'.

5   Experimental Results

So far, we have shown that our estimators exhibit quantitative robustness properties, but have not yet provided an indication of which would be more robust in practice. In this section, we present the results of an experiment in which the estimators are compared using the formal notion of *infinitesimal robustness* [7, 8, 19].

Infinitesimal robustness measures how small changes on the data set X affect the results of an estimator. In particular, if an estimator of location is differentiable and the derivative is bounded, small changes on the data can only amount to small changes on the estimate of location, and thus the estimator would be considered to be infinitesimally robust.

Let Δ_q be a distribution where the value $q \in \Re^D$ occurs with probability 1. Let P be another distribution, and consider the mixture of P with Δ_q, with elements drawn from P with probability $(1-\epsilon) > 0$, and from Δ_q with probability $\epsilon > 0$. We can write the resulting distribution as

$$P_{q,\epsilon} = (1 - \epsilon)P + \epsilon\Delta_q. \tag{1}$$

Mixture $P_{q,\epsilon}$ resembles a contaminated P where ϵ is a proportion of noise, all of it located at the point q. The relative influence on functional $T(P)$ of the replacement of P by $P_{q,\epsilon}$ is $[T(P_{q,\epsilon}) - T(P)]/\epsilon$, and the *influence function* of T at P is

$$IF(q) = \lim_{\epsilon \to 0^+} \frac{T(P_{q,\epsilon}) - T(P)}{\epsilon}.$$

A bounded influence function would indicate infinitesimal robustness.

In our experiments, we measured empirically the relative influence of estimators LMS_S$(P_{q,\epsilon})$ and LMS_P$(P_{q,\epsilon})$ with respect to LMS_D(P), capturing both the variation due to $\Delta(q)$ and due to the randomization of RS and RP. We chose P to be the normal distribution in 2 dimensions $N_{\mu,\Sigma}$, with μ as the origin and correlation matrix $\Sigma = I_2$ as the identity matrix. For Δ_q, we used the point $q = (400, 400)$. For given values of n and ϵ, we generated data sets X_i with n items drawn from the mixture. For the smaller values of n, we computed LMS_D(X_i) directly in $O(n^2)$ time; however, for large n the computational cost led us to use the origin (the distribution center for P) as an estimate.

Table 1 shows results for 24 different data sets generated from the mixture, in which the differences between LMS_D(X_i) and both LMS_S(X_i) and LMS_P(X_i) were analyzed. 95% confidence intervals for $E[$LMS_S$(X_i)]$ and $E[$LMS_P$(X_i)]$ are shown, estimated by executing the randomized algorithm 10 times for each data set. For reference, we also computed the sample mean AVG. For comparison purposes, both RS and RP were run with $r = u = \sqrt{n}$.

The experiment confirms the robustness of LMS_S(X) and LMS_P(X), even when 20% of the points were outliers drawn from $\Delta(q)$. The randomized nature of both is evident from the decrease in variation at the sample size increases. However, RP was considerably more accurate than RS over all sample sizes, confirming the theoretical advantages proven in the previous section.

					$\epsilon = 0.2$		
n	$\sqrt{n}$	ϵn	Data Set	LMS(X)	AVG(X)	$E[\text{LMS\_S}(X)] - \text{LMS}(X)$	$E[\text{LMS\_P}(X)] - \text{LMS}(X)$
100	10	20	X_1	(0.23,0.2)	(82.5,86.2)	(0.3±0.3,0.4 ± 0.8)	(0.19±0.14,0.07±0.08)
100	10	20	X_2	(-0.14,-0.49)	(100.5,99.9)	(0.4±0.3,0.1 ± 0.07)	(0.0±0.12,0.06±0.05)
900	30	180	X_3	(-0.08,0.06)	(88.2,89.3)	(0.3±0.3,0.3 ± 0.3)	(0.0±0.08,-0.01±0.02)
900	30	180	X_4	(-0.21,-0.0)	(76.8,76.8)	(0.2±0.4,0.2 ± 0.07)	(0.0±0.08,-0.05±0.05)
3,600	60	720	X_5	(0.04,-0.05)	(99.3,98.2)	(0.06±0.3,0.1 ± 0.2)	(0.0±0.01,-0.05±0.02)
3,600	60	720	X_6	(0.0,-0.01)	(92.2,91.3)	(-0.04±0.18,0.1 ± 0.07)	(0.0±0.01,-0.05±0.02)
10,000	100	2,000	X_7	not available	(95.2,94.5)	-(0.08±0.15,0.8 ± 0.06)	(0.0±0.02,-0.05±0.03)
10,000	100	2,000	X_8	not available	(94.5,94.7)	(0.12±0.1,-0.08 ± 0.15)	(0.0±0.02,-0.05±0.02)
					$\epsilon = 0.1$		
n	$\sqrt{n}$	ϵn	Data Set	LMS(X)	AVG(X)	$E[\text{LMS\_S}(X)] - \text{LMS}(X)$	$E[\text{LMS\_P}(X)] - \text{LMS}(X)$
100	10	10	X_9	(-0.4,0.3)	(38.9,38.8)	(0.4±0.11,0.1 ± 0.1)	(0.04±0.04,0.1±0.11)
100	10	10	X_{10}	(0.3,0.19)	(37.5,38.1)	(0.8±0.13,0.1 ± 0.3)	(0.0±0.06,0.0±0.11)
900	30	90	X_{11}	(0.2,0.1)	(38.9,38.3)	(0.3±0.1,0.3 ± 0.3)	(0.0±0.09,0.03±0.08)
900	30	90	X_{12}	(0.13,0.2)	(37.8,37.8)	(0.8±0.7,0.2 ± 0.6)	(0.0±0.03,0.0±0.02)
3,600	60	360	X_{13}	(0.05,0.16)	(35.9,36.2)	(0.2±0.3,-0.2 ± 0.7)	(0.0±0.05,0.0±0.03)
3,600	60	3360	X_{14}	(0.1,-0.1)	(39.2,39.1)	(0.3±0.7,0.4 ± 0.9)	(-0.01±0.04,0.07±0.02)
10,000	100	1,000	X_{15}	not available	(39.2,39.0)	(0.13±0.2,-0.13 ± 0.2)	(-0.11±0.02,0.5±0.07)
10,000	100	1,000	X_{16}	not available	(38.7,38.3)	(-0.16±0.15,0.11.8 ± 0.15)	(0.0±0.02,0.04±0.03)
					$\epsilon = 0.01$		
n	$\sqrt{n}$	ϵn	Data Set	LMS(X)	AVG(X)	$E[\text{LMS\_S}(X)] - \text{LMS}(X)$	$E[\text{LMS\_P}(X)] - \text{LMS}(X)$
100	10	1	X_{17}	(0.01,-0.1)	(3.9,3.9)	(-0.31±0.1,-0.62 ± 0.1)	(0.10±0.14,0.07±0.05)
100	10	1	X_{18}	(-0.2,0.1)	(3.5,3.9)	(0.8±0.3,0.5 ± 0.2)	(0.0±0.01,0.7±0.5)
900	30	9	X_{19}	(-0.12,0.1)	(3.9,3.4)	(0.07±0.15,0.17 ± 0.21)	(0.03±0.08,-0.02±0.03)
900	30	9	X_{20}	(0.1,-0.21)	(3.8,3.8)	(0.05±0.2,0.14 ± 0.17)	(0.02±0.04,0.02±0.03)
3,600	60	36	X_{21}	(0.05,0.00)	(3.4,3.5)	(-0.03±0.24,-0.10 ± 0.16)	(0.0±0.03,0.04±0.02)
3,600	60	36	X_{22}	(-0.01,-0.03)	(3.5,3.6)	(-0.04±0.18,0.12 ± 0.16)	(0.0±0.03,-0.04±0.02)
10,000	100	100	X_{23}	not available	(3.9,3.5)	(-0.07±0.15,-0.15 ± 0.24)	(0.0±0.01,0.01±0.05)
10,000	100	100	X_{24}	not available	(3.5,3.6)	(-0.08±0.13,0.2 ± 0.25)	(0.0±0.02,0.05±0.02)

Table 1. Comparing infinitesimal robustness.

6   Robust Clustering Applications

Expectation Maximization (EM) [3, 15, 16] describes a large family of iterative methods that have been applied to many different contexts [9], including learning in probabilistic networks [12], and clustering. The clustering problem can be viewed as that of finding a model to explain the data (typically a mixture of exponential distributions) and an assignment of the items to classes. EM methods for clustering proceed by repeated alternation between estimating the assignment of items to classes (the expectation step), and finding the model that maximizes the Maximum Likelihood of the classified data (the maximization step).

Other clustering methods exist which greatly resemble EM methods. For example, if the components of the mixture are assumed to be multivariate normals with known and common covariance Σ, and the only parameter missing in each component is the location, then EM is extremely similar to Fuzzy-C-Means [4, 2]. Moreover, if the assignment of items to classes is assumed to be crisp, EM reduces to k-Means. Different metrics and measures have also been used, such as the finding of class representatives using the spatial median under the L_1 metric (k-C-L_1-Medians [1, 5]) and the Euclidean metric (k-D-Medians [5]).

Algorithm	$\hat{\mu}_j$	$\|\hat{\mu}_j - \mu_j\|$	$\sum \|\hat{\mu}_j - \mu_j\|$	CPU time
LMS_P	$\mu_1 = (13.3, 7.6, 7.0)$	0.2	0.8	6 sec
	$\mu_2 = (2.8, 7.0, 14.7)$	0.3		
	$\mu_3 = (16.5, 9.1, 15.0)$	0.3		
FUZZY-C-MEANS	$\mu_1 = (12.7, 8.0, 6.7)$	0.6	1.9	1.11 sec
	$\mu_2 = (3.0, 7.4, 14.1)$	0.64		
	$\mu_3 = (16.2, 9.6, 14.5)$	0.64		
k-MEANS	$\mu_1 = (12.7, 8.4, 6.3)$	1.07	2.83	96 sec
	$\mu_2 = (3.1, 7.8, 13.5)$	1.32		
	$\mu_3 = (16.3, 9.6, 14.8)$	0.44		
EM	$\mu_1 = (10.1, 9.7, 9.4)$	3.8	7.77	5 sec
	$\mu_2 = (2.7, 7.1, 14.4)$	0.14		
	$\mu_3 = (15.1, 8.6, 11.3)$	3.83		
k-D-MEDIANS	$\mu_1 = (12.8, 8.0, 6.9)$	0.44	0.97	0.7 sec
	$\mu_2 = (2.8, 7.1, 14.4)$	0.22		
	$\mu_3 = (16.6, 9.6, 14.8)$	0.31		
k-C-L1-MEDIANS	$\mu_1 = (7.8, 16.0, 6.0)$	10.17	16.8	0.4 sec
	$\mu_2 = (3.8, 7.1, 14.4)$	1.2		
	$\mu_3 = (15.0, 8.5, 9.8)$	5.44		

Table 2. Comparison of Expectation Maximization variants.

The randomized estimators of location LMS_S(X), LTS_S(X), LMS_P(X) and LTS_P(X) can all be used in the maximization step to compute a representative of each class. We would expect that any general clustering method which used these estimators, and in particular LMS_P(X) and LTS_P(X), to be robust in comparison with the alternatives mentioned above. For comparison of the quality of clustering in the presence of noise, we implemented k-MEANS, EM, k-D-MEDIANS, k-C-L1-MEDIANS, FUZZY-C-MEANS, and a new algorithm using estimator LMS_P(X), which we simply refer to as LMS_P.

The data is drawn from a mixture of 3-dimensional normal distributions with noise. Data was generated with the form $p(x) = \pi_1 N_{\mu_1, \Sigma_1}(x) + \ldots + \pi_k N_{\mu_k, \Sigma_k}(x) + \pi_{k+1} U(x)$ where each component $N_{\mu_j, \Sigma_j}(x)$ is a multivariate normal distribution with mean μ_j and covariance matrix Σ_j. We used 20% noise, $k = 3$ and $\pi_j = .8/k$ for $j = 1, \ldots, k = 3$. The covariance matrices Σ_j were set to the identity so as to create data sets favorable to k-MEANS and EM.

We evaluated the quality of the clustering results by the sum of the norms between the original μ_j and the approximations $\hat{\mu}_j$ obtained for the algorithms. Data sets with $n = 2000$ were generated by selecting three points μ_j at random in $[0, 20.0] \times [0, 20.0] \times [0, 20.0]$. Table 2 shows the results for this dataset. It is clear that our randomized methods can be plugged into the Expectation Maximization framework to produce more robust clustering algorithms.

References

[1] P. S. Bradley, O. L. Mangasarian, and W. N. Street. Clustering via concave minimization. *Advances in Neural Information Processing Systems*, 9:368–374, 1997.
[2] V. Cherkassky and F. Muller. *Learning from Data — Concept, Theory and Methods*. John Wiley, NY, 1998.

[3] A. P. Dempster, N. M. Laird, and D. B. Rubin. Maximum likelihood from incomplete data via the EM algorithm. *J. Royal Statistical Society B*, 39:1–38, 1977.

[4] V. Estivill-Castro. Hybrid genetic algorithms are better for spatial clustering. In *Proc. 6th Pacific Rim Intern. Conf. Artificial Intelligence PRICAI 2000*, Melbourne, 2000. Springer-Verlag Lecture Notes in AI, to appear.

[5] V. Estivill-Castro and J. Yang. A fast and robust general purpose clustering algorithm. In *Proc. 6th Pacific Rim Intern. Conf. Artificial Intelligence PRICAI 2000*, Melbourne, 2000. Springer-Verlag Lecture Notes in AI, to appear.

[6] U. Fayyad and R. Uthurusamy. Data mining and knowledge discovery in databases. *Comm. ACM*, 39(11):24–26, Nov. 1996. Special issue on Data Mining.

[7] F. R. Hampel, E. M. Ronchetti, P. J. Rousseeuw, and W. A. Stahel. *Robust Statistics: The Approach Based on Influence Functions*. John Wiley, NY, 1986.

[8] P. J. Huber. *Robust Regression*. John Wiley, NY, 1981.

[9] M. I. Jordan and R. I. Jacobs. Supervised learning and divide-and-conquer: a statistical approach. In *Proc. 10th Intern. Conf. Machine Learning*, San Mateo, CA, Morgan Kaufmann, pp. 159–166, 1993.

[10] K. Koperski, J. Han, and J. Adhikary. Mining knowledge in geographical data. *Communications of the ACM*, to appear.

[11] R. Motwani and P. Raghavan. *Randomized Algorithms*. Cambridge University Press, UK, 1995.

[12] N. J. Nilsson. *Artificial Intelligence – A New Synthesis*. Morgan Kaufmann, 1998.

[13] E. M. Reingold, J. Nievergelt, and N. Deo. *Combinatorial Algorithms: Theory and Practice*. Prentice-Hall, Englewood Cliffs, NJ, 1977.

[14] P. J. Rousseeuw and A. M. Leroy. *Robust Regression and Outlier Detection*. John Wiley, NY, 1987.

[15] M. A. Tanner. *Tools for Statistical Inference*. Springer-Verlag, NY, USA, 1993.

[16] D. M. Titterington, A. F. M. Smith, and U. E. Makov. *Statistical Analysis of Finite Mixture Distributions*. John Wiley, UK, 1985.

[17] D. E. Tyler. Some issues in the robust estimation of multivariate location and scatter. In W. Stahel and S. Weisberg, eds., *Directions in Robust Statistics and Diagnostics — Part II*, Springer-Verlag, Berlin, pp. 327–336, 1991.

[18] C. S. Wallace. Intrinsic classfication of spatially correlated data. *The Computer Journal*, 41(8):602–611, 1998.

[19] R. R. Wilcox. *Introduction to Robust Estimation and Hypothesis Testing*. Academic Press, San Diego, CA, 1997.

Rough Sets in Spatio-temporal Data Mining*

Thomas Bittner

Centre de recherche en geomatique,
Laval University, Quebec, Canada
Thomas.Bittner@scg.ulaval.ca

Abstract. In this paper I define spatio-temporal regions as pairs consisting of a spatial and a temporal component and I define topological relations between them. Using the notion of rough sets I define approximations of spatio-temporal regions and relations between those approximations. Based on relations between approximated spatio-temporal regions configurations of spatio-temporal objects can be characterized even if only approximate descriptions of the objects forming them are available.

1 Introduction

Rough set theory [Paw82] provides a way of approximating subsets of a set when the set is equipped with a partition or equivalence relation. Rough sets were extensively used in the context of Data Mining, e.g., [Lin95, LC97]. So far, however, they were used mainly in non spatio-temporal contexts, for example, in order to classify and analyze phenomena, like diseases, given a finite number of observations or symptoms, e.g., [NSR92, BNSNT+95]. It is the purpose of this paper to apply rough sets in a spatio-temporal context, i.e., to describe and classify (configurations of) spatio-temporal objects.

An important task in spatio-temporal data mining is to discover characteristic configurations of spatial objects. Characterizing spatial configurations is important, for example, in order to retrieve your new 'ideal' home from a property database such that it has access to a highway, is located by the shore of a lake, within a beautiful forest, and far away from the next nuclear power station. Another important task is to find classes of configurations that characterize molecules like amino-acids and proteins [GFA93]. Spatio-*temporal* relations are often important to identify causal relationships between events in which are spatio-temporal objects are involved: In order to interact with each other things often need to be at the same place at the same time.

There are three major aspects characterizing *spatio-temporal objects*: (1) Aspects characterizing *what* they are, e.g., the class of things they belong to; (2) Aspects characterizing *where* they are, i.e., their spatial location; (3) Aspects characterizing *when* they existed and when they have been, are, or will be where, i.e., their temporal location. Between spatio-temporal objects hold *spatio-temporal relations* such as 'being in the same place at the same time', or 'having been in a place before something else'. Sets of spatio-temporal objects form *spatio-spatial configurations* that are characterized by sets

---

* The financial support from the Canadian GEOID network is gratefully acknowledged.

J.F. Roddick and K. Hornsby (Eds.): TSDM 2000, LNAI 2007, pp. 89–104, 2001.

of spatial, temporal, and spatio-temporal relations that hold between objects forming the configuration. In this paper I concentrate on topological spatio-temporal relations (like 'being in the same place at the same time'). Topological relations between regions of space and time play a major role in characterizing spatial and temporal configurations [EFG97, All83].

Today the classification of spatio-temporal configurations is based on relations between objects and on relations between the spatio-temporal regions they occupy. Unfortunately, it is often impossible to identify the region of space and time those objects exactly occupy, i.e., the exact location of spatio-temporal objects is often *indeterminate* [BF95]. [Bit99] argued that often *approximate* location of spatial objects is known. The notion of approximate location is based on the notion of rough sets, i.e., the approximation of (exact) location with respect to a regional partition of space and time. In this paper I discuss how rough sets can be used in order to describe approximate location in space and time and how to derive possible relations between objects given their approximations.

This paper is structured as follows. In Section 2 I define the notions of spatio-temporal object, location, region, and the relationships between them. I define topological relations between spatio-temporal regions in Section 3. The notion of a rough set is used in Section 4 in order to approximate spatio-temporal regions with respect to regional partitions of space and time. In Section 5 binary topological relations between those approximations are defined. These relations can be used to characterize configurations of spatio-temporal objects even if we know only their approximate location. In Section 6 the conclusions are given.

2   Location of Spatio-temporal Objects

Every spatio-temporal object, o, is located in[1] a unique region of time, $t(o)$, bounded by the begin and the end of its existence. In every moment of time a spatio-temporal object is exactly located in a single region, x^s, of space [CV95]. This region is the exact or precise spatial location of o at the time point t, i.e., $x^s = r_t(o)$ at t. Spatio-temporal wholes have temporal parts, which are located in parts of the temporal regions occupied by their wholes[2]. Consider, for example, the region of time, x^t, where the object, o, is located temporally, while being spatially located in the region x^s. If x^t is a maximal connected temporal region, i.e., o was once spatially located in x^s for a while, left and never came back, then x^t is bounded by the time instances (points) t_1 and t_2. Since time is a totally ordered set of time points (the set of all possible boundaries of time intervals) forming a directed one-dimensional space [Gea66], we have $t_1 < t_2$. In this paper time is modeled as a one dimensional directed line and space is modeled as a

---

[1]   We say that the object x is located *in* the (spatial, temporal or spatio-temporal) region y in order to stress the exact fit of object and region (the object matches the region). It is important to distinguish the exact match from the case of an object being located *within* a region which intuitive meaning allows the region to be bigger than the object and the case of the object *covering* a region which intuitively implies the region to be smaller than the object.

[2]   Notice that this implies a four dimensional ontology of spatio-temporal objects [Sim87]

2-dimensional plane. In the remainder I concentrate on *regions* of time and space and topological relations between them.

Spatio-temporal objects may be at rest, i.e., being located in the same region of space for a period of time, or they may change, i.e., being located in different regions of space at each moment of time[3]. Spatial change may be continuous, i.e., regions of consecutive moments of time are topologically close as in the case of change of bona-fide objects [SV97] like cars, planets, and human beings, or discontinuous, as (sometimes) in the case of change of fiat objects [SV97] like land property.

Consider an object at rest. I assume that the exact region of a spatio-temporal object, o, has always a corresponding time interval[4], x^t, which is bounded by the moment of time, t_1, where o 'stopped' at x^s and the moment, t_2, of time where o 'leaves' x^s (or the current moment of time if o is currently resting at x^s), i.e., $\forall t \in x^t : r_t(o) = x^s$. I define the *spatio-temporal region* of the *resting* object o as a pair, $r^r_{st}(o) = (x^t, x^s)$. The region x^t is a part of the exact temporal region of o, i.e., $P(x^t, t(o))$.

Spatial change causes spatio-temporal objects to be located in different regions of space at different moments of time. Consider a changing (moving, growing, shrinking, ...) spatio-temporal object, o, within the time interval $x^t = [t_b, t_e]$. Let x^{sm} be the sum, $\vee$, of the regions in which o was located during x^t, i.e., $x^{m_s} = \bigvee \{r_t(o) \mid t_b \le t \le t_e\}$. I define the *spatio-temporal region* of the *spatially changing* object o as a pair, $r^c_{st}(o) = (x^t, x^{sm})$. Notice, that doing this we do not know anymore, where exactly o is located during x^t. It can be everywhere within x^{sm}, but it cannot be somewhere else. In the special case of continuous movement the region x^{sm} can be thought of as the *path* of the object's movement during x^t. In the remainder I will use the metaphor 'path of change during x^t' in order to refer to the sum of spatial regions of a spatially changing object during the interval x^t.

3   Binary Topological Relations between Spatio-temporal Objects

Binary topological relations between regions such as overlap, contained/containing, disjoint, are well known in the spatial reasoning community, e.g., [EF91, RCC92]. Recently, [BS00] proposed a specific style that allows to define binary topological relations between regions exclusively based on constraints regarding the outcome of the meet (intersection) operation, denoted by $\wedge$, between (one and two dimensional) regions. This is critical for the generalization of these relations to the approximation case in Section 5. In this section I shortly review those definitions based on [BS00] and apply them to temporal and spatio-temporal regions afterwards.

3.1   Relations between Regions of (2D) Space

Given two regions x and y the boundary insensitive binary topological relation (RCC5 relations [RCC92]) between them can be determined by considering the triple of boolean values [BS00]:

$$(x \wedge y \ne \bot, \ x \wedge y = x, \ x \wedge y = y).$$

---

[3] This implies an ontology of absolute space and time, i.e., regions do not change.

[4] A maximal connected region of time.

The formula $x \wedge y \neq \bot$ is true if the intersection of y and y is not the empty region; The formula $x \wedge y = x$ is true if the intersection of x and y is identical to x; The formula $x \wedge y = y$ is true if the intersection of x and y is identical to y. The correspondence between such triples of boolean values and the RCC5 classification is given in the table below. Possible geometric interpretations are given in Figure 1 [BS00].

$x \wedge y \neq \bot$	$x \wedge y = x$	$x \wedge y = y$	RCC5
F	F	F	DR
T	F	F	PO
T	T	F	PP
T	F	T	PPi
T	T	T	EQ

The set of triples is partially ordered by defining $(a_1, a_2, a_3) \leq (b_1, b_2, b_3)$ iff $a_i \leq b_i$ for $i = 1, 2, 3$, where the boolean values are ordered by $\mathsf{F} < \mathsf{T}$. [BS00] refer to the Hasse diagram of the partially ordered set (The right diagram in Figure 1.) as the RCC5 *lattice*.

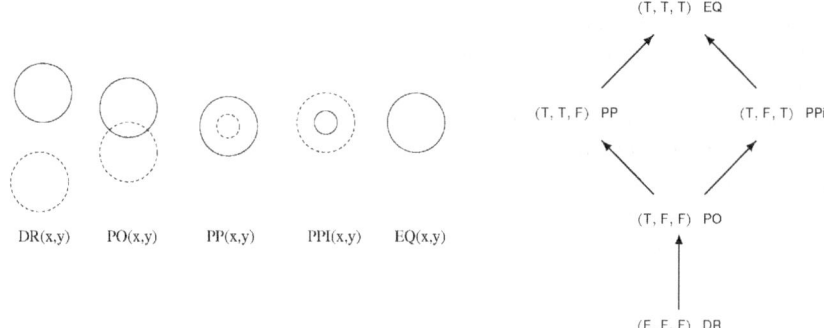

Fig. 1. RCC5 relations and RCC5 lattice

3.2   Relations between Temporal Regions

Consider the maximal connected one dimensional regions, x and y, i.e., intervals. Boundary insensitive topological relation between *intervals* x and y on a *directed* line (RCC$_1^9$ relations) can be determined by considering the triple of values belonging to the set
$\{\mathsf{FLO}, \mathsf{FLI}, \mathsf{T}, \mathsf{FRI}, \mathsf{FRO}\}$:

$$(x \wedge y \not\sim \bot, \ x \wedge y \sim x, \ x \wedge y \sim y)$$

where

$$x \wedge y \not\sim \bot = \begin{cases} \mathsf{FLO} & \text{if } x \wedge y = \bot \text{ and } x \ll y \\ \mathsf{FRO} & \text{if } x \wedge y = \bot \text{ and } x \gg y \\ T & \text{if } x \wedge y \neq \bot \end{cases}$$

where

$$x \wedge y \sim x = \begin{cases} \text{FLO} & \text{if } x \wedge y \neq x \text{ and } x \wedge y \neq y \text{ and } x \ll y \\ \text{FLI} & \text{if } x \wedge y \neq x \text{ and } x \wedge y = y \text{ and } x \ll y \\ \text{FRO} & \text{if } x \wedge y \neq x \text{ and } x \wedge y \neq y \text{ and } x \gg y \\ \text{FRI} & \text{if } x \wedge y \neq x \text{ and } x \wedge y = y \text{ and } x \gg y \\ \text{T} & \text{if } x \wedge y = x \end{cases}$$

and where

$$x \wedge y \sim y = \begin{cases} \text{FLO} & \text{if } x \wedge y \neq y \text{ and } x \wedge y \neq x \text{ and } x \ll y \\ \text{FLI} & \text{if } x \wedge y \neq y \text{ and } x \wedge y = x \text{ and } y \gg x \\ \text{FRO} & \text{if } x \wedge y \neq y \text{ and } x \wedge y \neq x \text{ and } x \gg y \\ \text{FRI} & \text{if } x \wedge y \neq y \text{ and } x \wedge y = x \text{ and } y \ll x \\ \text{T} & \text{if } x \wedge y = y \end{cases}$$

with

$$x \ll y = \begin{cases} T \text{ if } L(x) \wedge L(y) = L(x) \text{ and } L(x) \wedge L(y) \neq L(y) \\ F \text{ otherwise} \end{cases}$$

$$x \gg y = \begin{cases} T \text{ if } R(x) \wedge R(y) = R(x) \text{ and } R(x) \wedge R(y) \neq R(y) \\ F \text{ otherwise} \end{cases}$$

$L(x)$ $(R(y))$ is the one dimensional region occupying the whole line left (right)[5] of x. The intuition behind $x \wedge y \sim x = \text{FLO}$ ($x \wedge y \sim x = \text{FRO}$) is that "$x \wedge y = x$ is false because of parts of x 'sticking out' to the left (right) of y". The intuition behind $x \wedge y \sim y = \text{FLI}$ ($x \wedge y \sim y = \text{FRI}$) is that "$x \wedge y = y$ is false because of parts of y 'sticking out' to the right (left) x".

The triples formally describe jointly exhaustive relations under the assumption that x and y are intervals in a one dimensional directed space. The correspondence between the triples and the boundary insensitive relations between intervals is given in the table below. Possible geometric interpretations of the defined relations are given in Figure 2.

$x \wedge y \not\sim \bot$	$x \wedge y \sim x$	$x \wedge y \sim y$	RCC_1^9
FLO	FLO	FLO	DRL
FRO	FRO	FRO	DRR
T	FLO	FLO	POL
T	FRO	FRO	POR
T	T	FLI	PPL
T	T	FRI	PPR
T	FLI	T	PPiL
T	FRI	T	PPiR
T	T	T	EQ

For example. The relation $\text{DRL}(x, y)$ holds if x and y do not overlap and x is left of y; $\text{POL}(x, y)$ holds if x and y partly overlap and the non overlapping parts of x are left of y; $\text{PPL}(x, y)$ holds if x is contained in y but x does not cover the very right parts of

---

[5] I use the spatial metaphor of a line extending from the left to the right rather than the time-line extending from the past to the future in order to focus on the aspects of the time-line as a one-dimensional directed space. Time itself is much more difficult.

y; PPiL(x, y) holds if y is a part of x and there are parts of x sticking out to the left of y; PPR(x, y) holds if x is a part of y and x does not cover the very left parts of y; PPiR(x, y) holds if y is a part of x and there are parts of x sticking out to the right of y.

Assuming the ordering $FLO < FLI < \mathsf{T} < FRI < FRO$ a lattice is formed, which has (FLO, FLO, FLO) as minimal element and (FRO, FRO, FRO) as maximal element. The ordering is indicated by the arrows in Figure 2.

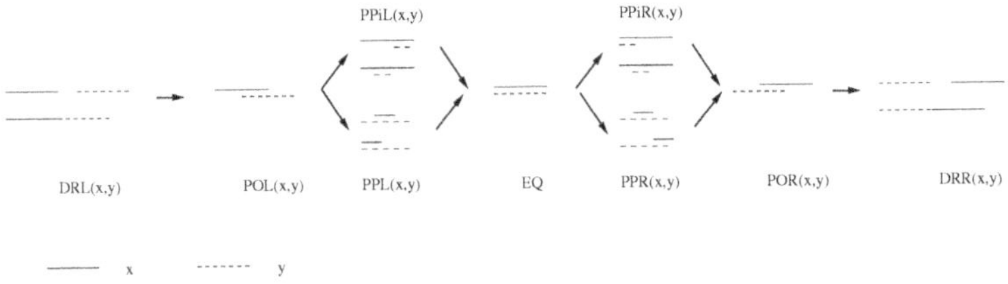

Fig. 2. Possible geometric interpretations of the RCC$_1^9$ relations.

3.3   Relations between Spatio-temporal Regions

Let o_1 and o_2 be two spatio-temporal objects at rest with spatio-temporal location $r_{st}^r(o_1) = (x^t, x^s)$ and $r_{st}^r(o_2) = (y^t, y^s)$, where x^t and y^t are time intervals, i.e., maximally connected temporal regions, and x^s and y^s are arbitrary, possibly scattered, 2D regions. The spatio-temporal relation between (x^t, x^s) and (y^t, y^s) can be described using the following pair of triples:

$$((x^t \wedge y^t \not\sim \bot, \ x^t \wedge y^t \sim x^t, \ x^t \wedge y^t \sim y^t),$$
$$(x^s \wedge y^s \neq \bot, \ x^s \wedge y^s = x^s, \ x^s \wedge y^s = y^s))$$

The relationship between those pairs of triples and spatio-temporal relations is given in the following table:

$x \wedge y \not\sim \bot$	$x \wedge y \sim x$	$x \wedge y \sim y$	$x \wedge y \neq \bot$	$x \wedge y = x$	$x \wedge y = y$	(RCC$_1^9$,RCC5)
FLO	FLO	FLO	F	F	F	(DRL,DR)
FLO	FLO	FLO	T	F	F	(DRL,PO)
$\cdots$	$\cdots$	$\cdots$	$\cdots$	$\cdots$	$\cdots$	$\cdots$
T	FLO	FLO	T	F	F	(POL,PO)
T	FLO	FLO	T	T	F	(POL,PP)
T	FLO	FLO	T	F	T	(POL,PPi)
T	FLO	FLO	T	T	T	(POL,EQ)
T	T	FLI	F	F	F	(PPL,DR)
T	T	FLI	T	F	F	(PPL,PO)
$\cdots$	$\cdots$	$\cdots$	$\cdots$	$\cdots$	$\cdots$	
T	T	T	T	T	T	(EQ,EQ)
FRO	FRO	FRO	F	F	F	(DRR,DR)
$\cdots$	$\cdots$	$\cdots$	$\cdots$	$\cdots$	$\cdots$	$\cdots$

The the Hasse diagram of the partially ordered set is called the $(\mathbf{RCC}_1^9, \mathbf{RCC5})$ *lattice*.

For example. The relation $(\mathbf{DRL}, \mathbf{PO})((x^t, x^s), (y^t, y^s))$ is interpreted as follows: The spatial regions x^s and y^s partially overlap and the relation between the time interval x^t when o_1 rested in x^s and the time interval y^t when o_2 rested in y^s is $\mathbf{DRL}$. The scenario, i.e., the sequence of events, could be described as: o_1 changes to location x^s and rests there during x^t. At some time in the future (o_1 has already left x^t), o_2 changes to y^s such that $\mathbf{PO}(x^s, y^s)$[6].

The relation $(\mathbf{POL}, \mathbf{PO})((x^t, x^s), (y^t, y^s))$ is interpreted as follows: The spatial regions x^s and y^s partially overlap and the relation between the time interval x^t when o_1 rested in x^s and the time interval y^t when o_2 rested in y^s is $\mathbf{POL}$. The scenario is: o_1 changes to its location x^s and rests there during x^t. While o_1 is resting in x^s, o_2 changes to y^s such that $\mathbf{PO}(x^s, y^s)$ holds. While o_2 is still resting in y^s, o_1 changes to another region. This new region may or may not overlap y^s.

Formally, for changing objects the same style of definition applies. Only the exact regions of o_1 and o_2, x^s and y^s during x^t and y^t, are replaced by the path of change x^{sm} and y^{sm}, of o_1 and o_2 during x^t and y^t. In this case we do not describe the relation between the location of rest of o_1 during x^t and the location of rest of o_2 during y^t, but the relation between the path of change of o_1 during x^t and the path of change of o_2 during y^t. The relation $(\mathbf{POL}, \mathbf{PO})((x^t, x^s), (y^t, y^s))$ is interpreted as follows: The path of change of o_1 during x^t and the path of change of o_2 during y^t do partially overlap and the relation $\mathbf{POL}(x^t, y^t)$ holds between the time intervals x^t and y^t. The interpretation of $\mathbf{POL}(x^t, y^t)$ is that we started monitoring the path of o_1 earlier than monitoring the path of o_2 and finished monitoring the path of o_1 earlier than monitoring the path of o_2.

4   Rough Approximations

Rough set theory [Paw82] provides a way of approximating subsets of a set when the set is equipped with a partition or equivalence relation. Given a set X with a partition $\{a_i \mid i \in \mathcal{I}\}$, an arbitrary subset $b \subseteq X$ can be approximated by a function $\varphi_b : \mathcal{I} \to \{\mathbf{fo}, \mathbf{po}, \mathbf{no}\}$. The value of $\varphi_b(i)$ is defined to be $\mathbf{fo}$ if $a_i \subseteq b$, it is $\mathbf{no}$ if $a_i \cap b = \varnothing$, and otherwise the value is $\mathbf{po}$. The three values $\mathbf{fo}$, $\mathbf{po}$, and $\mathbf{no}$ stand respectively for 'full overlap', 'partial overlap' and 'no overlap'; they measure the extent to which b overlaps the elements of the partition of X.

4.1   Approximating Spatial and Temporal Regions

[BS00] showed that regions of space and time can be described by specifying how they relate to a partition of space and time into cells which may share boundaries but which do not overlap. A region can then be described by giving the relationship between the region and each cell. Suppose a space $\mathcal{S}$ of precise regions. By imposing a partition, G, on $\mathcal{S}$ we can approximate elements of $\mathcal{S}$ by elements of Ω_3^G. That is, we approximate regions in $\mathcal{S}$ by functions from G to the set $\Omega_3 = \{\mathbf{fo}, \mathbf{po}, \mathbf{no}\}$. The function which

---

[6] Remember, regions do not change.

assigns to each region $x \in \mathcal{S}$ its approximation is denoted $\alpha_3 : \mathcal{S} \to \Omega_3^G$. The value of $(\alpha_3 \, x) \, g$ is fo if x covers all the of the cell g, it is po if x covers some but not all of the interior of g, and it is no if there is no overlap between x and g.

Each approximate region $X \in \Omega_3^G$ stands for a set of precise regions, i.e., all those precise regions having the approximation X. This set which will be denoted $[\![X]\!]$ provides a semantics for approximate regions: $[\![X]\!] = \{x \in \mathcal{S} \mid \alpha_3 x = X\}$ [BS00].

4.2   The Meet Operation

The domain of regions is equipped with a meet operation interpreted as the intersection of regions. In the domain of approximation functions the meet operation between regions is approximated by pairs of greatest minimal, $\underline{\wedge}$, and least maximal, $\overline{\wedge}$, meet operations on approximation mappings [BS98]. Consider the operations $\underline{\wedge}$ and $\overline{\wedge}$ on the set $\Omega_3 = \{\text{fo}, \text{po}, \text{no}\}$ that are defined as follows:

$\underline{\wedge}$	no	po	fo
no	no	no	no
po	no	no	po
fo	no	po	fo

$\overline{\wedge}$	no	po	fo
no	no	no	no
po	no	po	po
fo	no	po	fo

These operations extend to elements of Ω_3^G (i.e. the set of functions from G to Ω_3) by

$$(X \underline{\wedge} Y)g = (Xg) \underline{\wedge} (Yg)$$

and similarly for $\overline{\wedge}$.

4.3   Approximating Spatio-temporal Regions

Spatio-temporal regions are pairs, (x^t, x^s), consisting of a spatial component, x^s, and a temporal component, x^t. Both components can be approximated separately by approximation functions, X^s and X^t with respect to partitions G_T and G_S of time and space, as described above. Consequently, an approximate spatio-temporal regions is a pair (X^t, X^s). Each approximate spatio-temporal region $(X^t, X^s) \in \Omega^{G_T} \times \Omega^{G_S}$ stands for a set of precise spatio-temporal regions, i.e., all those precise regions having the approximation (X^t, X^s). This set which will be denoted $[\![(X^t, X^s)]\!]$ provides a semantics for approximate spatio-temporal regions:

$$[\![(X^t, X^s)]\!] = \{(x^t, x^s) \in \mathcal{T} \times \mathcal{S} \mid \alpha \, x^t = X^t \text{ and } \alpha \, x^s = X^s\},$$

where $\mathcal{T}$ denotes the set of regions of the time-line and $\mathcal{S}$ denotes the regions of the plane. The greatest minimal and least maximal meet operations between approximations of spatial and temporal regions generalize in the natural way to approximations of spatio-temporal regions:

$$(X^t, X^s) \underline{\wedge} (Y^t, Y^s) = (X^t \underline{\wedge} Y^t, X^s \underline{\wedge} Y^s)$$

and similarly for $\overline{\wedge}$.

5   Approximating Binary Topological Relations

I discussed above the importance of qualitative spatial relation for the description of spatial configurations. In this section I define approximate topological relations between approximations of spatial, temporal, and spatio-temporal regions. In this context I apply the specific style of definitions discussed in Section 3. Relations between approximations of spatial have been discussed separately in [BS00]. This will be reviewed and then applied to relations between approximated temporal and spatio-temporal regions.

In order to define relations between approximations of spatial regions and temporal intervals I firstly pursue the syntactic approach: (1) I replace in the definitions of relations between spatial regions and temporal intervals the (variables ranging over) regions by (variables ranging over) approximations, e.g., I replace $x \in S$ by $X \in \Omega_3^G$; and (2) I replace the meet operation between regions by the greatest minimal and least maximal operations between approximations, i.e., I replace $\wedge$ by $\underline{\wedge}$ and $\overline{\wedge}$. Secondly, I check whether the syntactically generated pairs of relations constrain the appropriate set of relations between the approximated regions (the semantic approach).

5.1   Relations between Approximations of Spatial Regions

The above formulation of the RCC5 relations can be extended to approximate regions. One way to do this is to replace the operation $\wedge$ with an appropriate operation for approximate regions. If X and Y are approximate regions (i.e. functions from G to Ω_3) we can consider the two triples of Boolean values [BS00]:

$$(X \underline{\wedge} Y \neq \bot, \ X \underline{\wedge} Y = X, \ X \underline{\wedge} Y = Y),$$
$$(X \overline{\wedge} Y \neq \bot, \ X \overline{\wedge} Y = X, \ X \overline{\wedge} Y = Y).$$

In the context of approximate regions, the bottom element, $\bot$, is the function from G to Ω_3 which takes the value no for every element of G. Each of the above triples provides an RCC5 relation, so the relation between X and Y can be measured by a pair of RCC5 relations. These relations will be denoted by $\underline{R}(X,Y)$ and $\overline{R}(X,Y)$. The pairs $(\underline{R}(X,Y), \overline{R}(X,Y))$ which can occur are all pairs (a, b) where $a \leq b$ with the exception of $(\mathsf{PP}, \mathsf{EQ})$ and $(\mathsf{PPi}, \mathsf{EQ})$ [BS00].

Consider the ordering of the RCC5 lattice. The relation $\underline{R}(X,Y)$ is the minimal relation and the relation $\overline{R}(X,Y)$ is the maximal relation that can hold between $x \in [\![X]\!]$ and $y \in [\![Y]\!]$. For all relations R, with $\underline{R}(X,Y) \leq R \leq \overline{R}(X,Y)$ there are $x \in [\![X]\!]$ and $y \in [\![Y]\!]$ such that $R(x, y)$ [BS00].

5.2   Syntactic Generalization of Relations between Temporal Intervals

In order to generalize the above formulation of RCC_1^9 relations to relations between approximations of temporal intervals we need to define operations $X \ll Y$ and $X \gg Y$ corresponding to operations $x \ll y$ and $x \gg y$. The behavior of $X \ll Y$ is shown in Figure 3. Formally we define $X \ll Y$ as

$$X \ll Y = \begin{cases} \mathsf{T} \ \text{if } L(X) \overline{\wedge} L(Y) = L(X) \text{and } L(X) \overline{\wedge} L(Y) \neq L(Y) \\ \mathsf{M} \ \text{if } L(X) \overline{\wedge} L(Y) = L(X) \text{ and } L(X) \overline{\wedge} L(Y) = L(Y) \text{ and} \\ \quad L(X) \underline{\wedge} L(Y) < L(X) \overline{\wedge} L(Y) \\ \mathsf{F} \ \text{otherwise} \end{cases}$$

and similarly $X \gg Y$ using $R(X)$ and $R(Y)$, where $L(X)$ yields the approximation of the part of the time-line left of $x \in [\![X]\!]$ and $R(Y)$ yields the approximation of the part of the time-line right of $y \in [\![Y]\!]$ respectively. Formally, L and R are defined as follows. Firstly, we define the complement operation $X' g_i = (X g_i)'$ with $\mathsf{no}' = \mathsf{fo}$, $\mathsf{po}' = \mathsf{po}$, and $\mathsf{fo}' = \mathsf{no}$. Assuming that partition cells g_i are numbered in increasing order in direction of the underlying space, we secondly define $L(X)$ and $R(Y)$ as:

$$(L(X) g_i) = \begin{cases} (X g_i)' & \text{if } i \leq \min\{k \\ & \mid (X g_k) \neq \mathsf{no}\} \\ \mathsf{no} & otherwise \end{cases} ; \quad (R(Y) g_i) = \begin{cases} (Y g_i)' & \text{if } i \geq \max\{k \\ & \mid (Y g_k) \neq \mathsf{no}\} \\ \mathsf{no} & otherwise \end{cases} .$$

$$X \ll Y = \mathsf{T} \qquad X \ll Y = \mathsf{T} \qquad X \ll Y = \mathsf{M} \qquad X \ll Y = \mathsf{F}$$

Fig. 3. The behavior of $X \ll Y$, where $x \in [\![X]\!]$ is above the time-line and $y \in [\![Y]\!]$ is below the time-line.

We need two more operations: $X \triangleright Y$ and $X \triangleleft Y$, where $X \triangleright Y = \mathsf{T}$ means that $x \in [\![X]\!]$ is contained in $y \in [\![Y]\!]$ and x does not cover the very right parts of y and $X \triangleleft Y = \mathsf{T}$ is interpreted as $x \in [\![X]\!]$ is contained in $y \in [\![y]\!]$ and x does not cover the very left parts of y. The behavior of $X \triangleright Y$ are shown in Figure 4. Formally we define a set $\Gamma(X, Y) = \{(R(X) \triangle Y) g_i \mid g_i \in G\}$, containing the elements of the co-domain of $(R(X) \triangle Y)$, and the operation

$$X \triangleright Y = \begin{cases} \mathsf{T} & \text{if } \mathsf{fo} \in \Gamma(X, Y) \text{ or } \{\mathsf{po}, \mathsf{po}\} \subseteq \Gamma(X, Y) \\ \mathsf{M} & \text{if } \mathsf{po} \in \Gamma(X, Y) \text{ and } \{\mathsf{po}, \mathsf{po}\} \not\subseteq \Gamma(X, Y) \\ \mathsf{F} & otherwise \end{cases} .$$

We define $X \triangleleft Y$ respectively by replacing $R(X)$ by $L(X)$ in the definition of $X \triangleright Y$.

$$X \triangleright Y = \mathsf{T} \qquad X \triangleright Y = \mathsf{T} \qquad X \triangleright Y = \mathsf{M} \qquad X \triangleright Y = \mathsf{F}$$

Fig. 4. The behavior of $X \triangleright Y$, where $x \in [\![X]\!]$ is above the time-line and $y \in [\![Y]\!]$ is below the time-line.

We are now able to generalize the above formulation of RCC_1^9 relations to relations between approximations. Let X and Y be boundary insensitive approximations of temporal intervals. We can consider the two triples of values:

$$((X \underline{\wedge} Y \not\sim \perp, \ X \underline{\wedge} Y \sim X, \ X \underline{\wedge} Y \sim Y),$$
$$(X \overline{\wedge} Y \not\sim \perp, \ X \overline{\wedge} Y \sim X, \ X \overline{\wedge} Y \sim Y)).$$

where

$$X \underline{\wedge} Y \not\sim \bot = \begin{cases} FLO & \text{if } X \underline{\wedge} Y = \bot \text{ and } (X \ll Y) \neq \mathsf{F} \text{ and } (X \ll Y) \geq (X \gg Y) \\ FRO & \text{if } X \underline{\wedge} Y = \bot \text{ and } (X \gg Y) \neq \mathsf{F} \text{ and } (X \gg Y) > (X \ll Y) \\ T & \text{if } X \underline{\wedge} Y \neq \bot \end{cases}$$

where

$$X \underline{\wedge} Y \sim X = \begin{cases} FLO & \text{if } X \underline{\wedge} Y \neq X \text{ and } X \underline{\wedge} Y \neq Y \text{ and } X \ll Y \neq \mathsf{F} \text{ and } X \ll Y \geq X \gg Y \\ FLI & \text{if } X \underline{\wedge} Y \neq X \text{ and } X \underline{\wedge} Y = Y \text{ and } leftCheck(X,Y) \\ FRO & \text{if } X \underline{\wedge} Y \neq X \text{ and } X \underline{\wedge} Y \neq Y \text{ and } X \gg Y \neq \mathsf{F} \text{ and } X \gg Y > X \ll Y \\ FRI & \text{if } X \underline{\wedge} Y \neq X \text{ and } X \underline{\wedge} Y = Y \text{ and } rightCheck(X,Y) \\ T & \text{if } X \underline{\wedge} Y = X \end{cases}$$

and where

$$X \underline{\wedge} Y \sim X = \begin{cases} FLO & \text{if } X \underline{\wedge} Y \neq Y \text{ and } X \underline{\wedge} Y \neq X \text{ and } X \ll Y \neq \mathsf{F} \text{ and } X \ll Y \geq X \gg Y \\ FLI & \text{if } X \underline{\wedge} Y \neq Y \text{ and } X \underline{\wedge} Y = X \text{ and } rightCheck(Y,X) \\ FRO & \text{if } X \underline{\wedge} Y \neq Y \text{ and } X \underline{\wedge} Y \neq X \text{ and } X \gg Y \neq \mathsf{F} \text{ and } X \gg Y > X \ll Y. \\ FRI & \text{if } X \underline{\wedge} Y \neq Y \text{ and } X \underline{\wedge} Y = X \text{ and } leftCheck(Y,X) \\ T & \text{if } X \underline{\wedge} Y = X \end{cases}$$

The functions $leftCheck(X,Y)$ and $rightCheck(X,Y)$ are defined as follows:

$$leftCheck(X,Y) = \begin{cases} T & \text{if } Y \triangleleft X = T \text{ or } (Y \triangleleft X = M \text{ and } Y \triangleright X = F) \\ F & \text{if } Y \triangleleft X \neq T \text{ and } Y \triangleright X = T \\ X \ll Y \neq \mathsf{F} \text{ and} & \text{otherwise} \\ X \ll Y \geq X \gg Y \end{cases},$$

$$rightCheck(X,Y) = \begin{cases} T & \text{if } Y \triangleright X = T \text{ or } (Y \triangleright X = M \text{ and } Y \triangleleft X = F) \\ F & \text{if } Y \triangleright X \neq T \text{ and } Y \triangleleft X = T \\ X \gg Y \neq \mathsf{F} \text{ and} & \text{otherwise} \\ X \gg Y > X \ll Y \end{cases}.$$

Both functions assume that $y \in [\![Y]\!]$ is contained in $x \in [\![X]\!]$. The behavior of $leftCheck$ is shown in Figure 5. The definitions of $X \overline{\wedge} Y \not\sim \bot$, $X \overline{\wedge} Y \sim Y$, and $X \overline{\wedge} Y \sim Y$ are obtained by replacing $\underline{\wedge}$ by $\overline{\wedge}$ in the definitions of $X \underline{\wedge} Y \not\sim \bot$, $X \underline{\wedge} Y \sim Y$, and $X \underline{\wedge} Y \sim Y$.

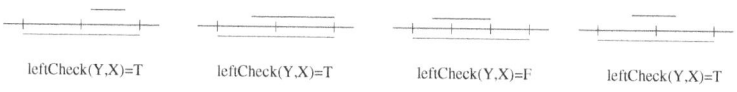

| leftCheck(Y,X)=T | leftCheck(Y,X)=T | leftCheck(Y,X)=F | leftCheck(Y,X)=T |

Fig. 5. The behavior of $leftCheck(Y,X)$, where $x \in [\![X]\!]$ is above the time-line and $y \in [\![Y]\!]$ is below the time-line.

Each of the above triples defines an RCC_1^9 relation, so the relation between X and Y can be measured by a pair of RCC_1^9 relations. These relations will be denoted by $\underline{R^9}(X,Y)$ and $\overline{R^9}(X,Y)$.

Theorem 1 *The pairs*

$$(\min\{\underline{R}^9(X,Y), \overline{R^9}(X,Y)\}, \max\{\underline{R}^9(X,Y), \overline{R^9}(X,Y)\})$$

that can occur are all pairs (a, b) *where* $a \leq b \leq$ EQ *and* EQ $\leq a \leq b$ *with the exception of* (PPL, EQ), (PPR, EQ), (PPiL, EQ), (PPiR, EQ), *and* (EQ, DRR).

Proof   The pairs (PPL, EQ), (PPR, EQ), (PPiL, EQ), (PPiR, EQ) cannot occur since RCC_1^9 relations are refinements of RCC5 relations and the pairs (PP, EQ) and (PPi, EQ) cannot occur in the RCC5 case [BS00]. The pair (EQ, DRR) cannot occur due to the non-symmetry of the underlying definitions. In order to generate all remaining pairs approximations of time intervals in regional partitions consisting of at least three elements need to be considered. A Haskell [Tho99] program generating all remaining pairs of relations between approximations with respect to a partition consisting of three intervals can be found at [Bit00b].   ▯

5.3   Semantic Generalization of Relations between Temporal Intervals

At the semantic level we consider how syntactically generated pairs, $(\underline{R}^9(X,Y),$ $\overline{R^9}(X,Y))$[7], relate to relations between the approximated regions $x \in \llbracket X \rrbracket$ and $y \in \llbracket Y \rrbracket$. The aim is that the syntactically generated pairs constrain the possible relations that can hold between the approximated intervals x and y [BS00]:

$$\{R \mid \underline{R}^9(X,Y) \leq R(X,Y) \leq \overline{R^9}(X,Y)\} = \{\rho(x,y) \mid x \in \llbracket X \rrbracket, y \in \llbracket Y \rrbracket\}$$

We proceed by considering all pairs containing the relation EQ. Consider configuration (a) in Figure 6, which represents the most indeterminate case. The syntactic approach described above yields the pair (DRL, EQ). Since in this kind of configuration the pair (DRL, EQ) is consistent with (EQ, DRR) and (DRL, EQ) was chosen arbitrarily, (DRL, EQ) is corrected syntactically to (DRL, DRR).

Consider configuration (b) in Figure 6. The syntactic approach yields the pair (DRL, EQ) which is not correct if x and y are intervals as depicted. Notice that the meet operations were originally defined for arbitrary regions not for one-dimensional intervals. Assuming $x \in \llbracket X \rrbracket$ and $y \in \llbracket Y \rrbracket$ to be (time) intervals the outcome of the minimal meet must not be empty. This needs to be taken into account in the definition of $\triangle$. Let X and Y be boundary insensitive approximations of time intervals:

$$(X \, g_i)(\triangle')(Y \, g_i) = \begin{cases} \text{PO} & \text{if } ((X \, g_i) = \text{PO and } (Y \, g_i) = \text{PO}) \text{ and} \\ & ((X \, g_{i-1}) \geq \text{PO and } (Y \, g_{i-1}) \geq \text{PO}) \text{ or} \\ & (X \, g_{i+1}) \geq \text{PO and } (Y \, g_{i+1}) \geq \text{PO})) \\ (X \, g_i) \triangle (Y \, g_i) & \text{otherwise} \end{cases}$$

Applying $(\triangle')$ to X and Y in Figure 6 (b) yields EQ as minimal relation. But (EQ, EQ) still does not characterize Figure 6 (b) correctly, since between $x \in \llbracket X \rrbracket$

---

[7] In the remainder of the paper I write $(\underline{R}^9(X,Y), \overline{R^9}(X,Y))$ instead of $(\min\{\underline{R}^9(X,Y), \overline{R^9}(X,Y)\}, \max\{\underline{R}^9(X,Y), \overline{R^9}(X,Y)\})$.

and $y \in [\![Y]\!]$ the relations $\{\mathsf{POL}(x,y), \mathsf{EQ}(x,y), \mathsf{POR}(x,y)\}$ can hold. Consider also Figure 6 (c) for which the operations defined above yield $(\mathsf{POL}, \mathsf{EQ})(X,Y)$, but the approximations X and Y are also consistent with $\mathsf{POR}(x,y)$ for $x \in [\![X]\!]$ and $y \in [\![Y]\!]$. Consequently, if $\max\{\underline{R}^9(X,Y), \overline{R}^9(X,Y)\} = \mathsf{EQ}$ and $\min\{\underline{R}^9(X,Y), \overline{R}^9(X,Y)\} \neq \mathsf{DRL}$ and the leftmost *or* the rightmost non-empty approximation values of X *and* Y have the value PO then the RCC_1^9 relation between X and Y is $(\mathsf{POL}, \mathsf{POR})(X,Y)$. This also applies to the configuration Figure 6 (d). The corrected relations are denoted $\underline{R}_c^9(X,Y)$ and $\overline{R}_c^9(X,Y)$.

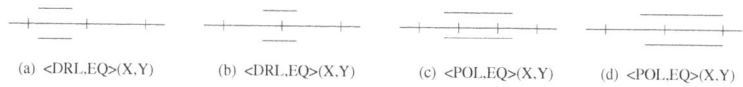

(a) <DRL,EQ>(X,Y)   (b) <DRL,EQ>(X,Y)   (c) <POL,EQ>(X,Y)   (d) <POL,EQ>(X,Y)

Fig. 6. Configurations characterized by pairs containing the relation $\mathsf{EQ}(X,Y)$, where $x \in [\![X]\!]$ is above the time-line and $y \in [\![Y]\!]$ is below the time-line.

Finally, consider the configuration (a) in Figure 7. Our definitions yield $(\mathsf{PPiL}, \mathsf{PPiL})(X,Y)$ but the approximations X and Y are also consistent with $\mathsf{PPiR}(x,y)$ but not with $\mathsf{EQ}(x,y)$ for $x \in [\![X]\!]$ and $y \in [\![Y]\!]$. The interval y can cover the very right part of x or the very left part of x or some part in the middle. The same holds if we switch X and Y (Figure 7 (b)): Our definitions yield $(\mathsf{PPL}, \mathsf{PPL})(X,Y)$ but the approximations X and Y are also consistent with $\mathsf{PPR}(x,y)$ but not with $\mathsf{EQ}(x,y)$. Consider Figure 7 (c) and (d). These cases are different: Assuming that x and y are intervals x can cover neither the very letf of y nor the very right of y. Consequently, the configuration is consistent with both $(\mathsf{PPL}, \mathsf{PPL})$ and $(\mathsf{PPR}, \mathsf{PPR})$ but in this cases it is o.k. to chose one, since x must cover parts in the middle of y.

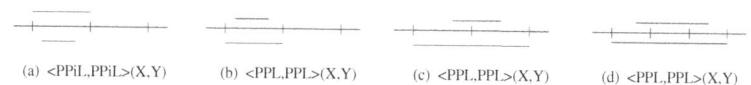

(a) <PPiL,PPiL>(X,Y)   (b) <PPL,PPL>(X,Y)   (c) <PPL,PPL>(X,Y)   (d) <PPL,PPL>(X,Y)

Fig. 7. Configurations characterized by pairs $(\mathsf{PPL}, \mathsf{PPL})$ or $(\mathsf{PPiL}, \mathsf{PPiL})$, where $x \in [\![X]\!]$ is above the time-line and $y \in [\![Y]\!]$ is below the time-line.

The cases depicted in Figure 7 (a) and (b) need to be handled separately. For them the theorem below does not hold. The problem disappears if boundary sensitive approximations are used. For all other case we state Theorem 2:

Theorem 2 *The relation $\underline{R}_c^9(X,Y)$ is the minimal relation and the relation $\overline{R}_c^9(X,Y)$ is the maximal relation that can hold between $x \in [\![X]\!]$ and $y \in [\![Y]\!]$. For all relations R, with $\underline{R}_c^9(X,Y) \leq R \leq \overline{R}_c^9(X,Y)$ there are $x \in [\![X]\!]$ and $y \in [\![Y]\!]$ such that $R(x,y)$.*

Proof RCC_1^9 relations are refinements of RCC5 relations. Figure 2 shows that the RCC_1^9 lattice can be separated into the left and the right RCC5 sub-lattices ($\mathsf{DRL} \leq$

$R \leq$ EQ and EQ $\leq R \leq$ DRR). Theorem 1 tells us that our syntactic procedure yields minimal and maximal relation pairs that either belong to the left RCC5 sub-lattice or the right RCC5 sublattice. It also tells us that the generated pairs are same pairs occuring in the RCC5 case. Consequently, with the exeption of the special cases discussed above, Theorem 2 of [BS00] applies, stating that the syntactic approach constrains the right set of relations. Consequently, what remains to show is that the theorem holds for the special cases: (DRL, DRR) and (POL, POR).

The case (DRL, DRR) occurs in configurations where the syntactic procedure yields (DRL, EQ), i.e., in configurations that are equivalent to the configuration in Figure 6 (a). DRL and DRR are trivially minimal and maximal and it is easy to verify that all relation ρ with DRL$(X,Y) \leq \rho(x,y) \leq$ DRR(X,Y) can actually occur for $x \in [\![X]\!]$ and $y \in [\![Y]\!]$.

The case (POL, POR) occurs in configurations where the syntactic procedure yields that $\max\{\underline{R}^9(X,Y), \overline{R}^9(X,Y)\}$ = EQ and $\min\{\underline{R}^9(X,Y), \overline{R}^9(X,Y)\} \neq$ DRL and that the leftmost or the rightmost non-empty approximation values of X and Y have the value PO, i.e., in configurations that are similar to the configuration in Figure 6 (b-d). It is easy to verify that exactly the relations ρ with POL$(X,Y) \leq \rho(x,y) \leq$ POR(X,Y) can actually occur for $x \in [\![X]\!]$ and $y \in [\![Y]\!]$.

5.4   Approximating Topological Relations between Spatio-temporal Objects

Based on relations between approximations of spatial regions and relations between approximations of temporal regions we now define relations between approximations of spatio-temporal regions. Let o_1 and o_2 be two spatio-temporal objects at rest with spatio-temporal location $r^c_{st}(o_1) = (x^t, x^s)$ and $r^c_{st}(o_2) = (y^t, y^s)$ with approximations (X^t, X^s) and (Y^t, Y^s). Consider the following structure:

$$(((X^t \underline{\wedge} Y^t \not\sim \perp, X^t \underline{\wedge} Y^t \sim X^t, X^t \underline{\wedge} Y^t \sim Y^t)$$
$$(X^s \underline{\wedge} Y^s \neq \perp, X^s \underline{\wedge} Y^s = X^s, X^s \underline{\wedge} Y^s = Y^s)),$$
$$((X^t \overline{\wedge} Y^t \not\sim \perp, X^t \overline{\wedge} Y^t \sim X^t, X^t \overline{\wedge} Y^t \sim Y^t),$$
$$(X^s \overline{\wedge} Y^s \neq \perp, X^s \overline{\wedge} Y^s = X^s, X^s \overline{\wedge} Y^s = Y^s)))$$

Each cpmponent of the above pair of pairs of triples defines a spatio-temporal relation, (RCC$^9_1$, RCC5). So the relation between (X^t, X^s) and (Y^t, Y^s) can be measured by a pair of spatio-temporal relations: $((\underline{R}^9(X,Y), \underline{R}(X,Y)), (\overline{R}^9(X,Y), \overline{R}(X,Y)))$. The pairs

$$((\min\{\underline{R}^9_c(X,Y), \overline{R}^9_c(X,Y)\}), \underline{R}), (\max\{\underline{R}^9_c(X,Y), \overline{R}^9_c(X,Y)\}, \overline{R}))$$

that can occur are exactly those that can occur in the separate treatment of approximations of RCC5 and RCC$^9_1$ relations.

Consequently, relations between approximations of spatio-temporal regions, (X^t, X^s) and (Y^t, Y^s), are represented by pairs of minimal and maximal spatio-temporal relations $((\underline{R}^9_c, \underline{R}), (\overline{R}^9_c, \overline{R}))$ such that $(\underline{R}^9_c, \underline{R})((X^s, X^t), (Y^t, Y^s))$ is the least spatio-temporal relation and $(\overline{R}^9_c, \overline{R})((X^s, X^t), (Y^t, Y^s))$ is the largest spatio-temporal relation that can hold between spatio-temporal regions $(x^t, x^s) \in [\![(X^t, X^s)]\!]$ and $(y^t, y^s) \in [\![(Y^t, Y^s)]\!]$.

6   Conclusions

In this paper I defined spatio-temporal regions as pairs consisting of a spatial and a temporal component. I defined topological relations between spatio-temporal regions based on topological relations between the spatial and temporal components. Approximations of spatio-temporal regions were defined using approximations of their spatial and temporal components. I defined topological relations between approximations of spatio-temporal regions based on a specific style that allows to define relations between spatio-temporal regions exclusively based on constraints on the outcome on the meet operation. The proposed framework can be used in order to describe spatial configurations based on approximate descriptions of spatio-temporal objects and relation between those approximations. Those approximate descriptions can be much easier obtained from observations of reality than exact descriptions.

The formalism discussed in this paper deals only with boundary insensitive topological relations between spatio-temporal regions. This can be easily extended to boundary sensitive relations using the formalisms proposed in [BS00] and [Bit00a].

References

[All83]   J.F. Allen. Maintaining knowledge about temporal intervals. *Communications of the ACM*, 26(11):832–843, 1983.

[BF95]   Peter Burrough and Andrew U. Frank, editors. *Geographic Objects with Indeterminate Boundaries*. GISDATA Series II. Taylor and Francis, London, 1995.

[Bit99]   T. Bittner. On ontology and epistemology of rough location. In *Spatial information theory - Cognitive and computational foundations of geographic information science, COSIT 99*, number 1661 in Lecture Notes in Computer Science, Hamburg, Germany, 1999. Springer Verlag.

[Bit00a]   T. Bittner. Approximate temporal reasoning. In *Workshop proceedings of the Seventeenth National Conference on Artificial Intelligence, AAAI 2000*, 2000.

[Bit00b]   T. Bittner. A Haskell program generating all possible relations between boundary insensitive approximations of time intervals. http://www.cs.queensu.ca/~bittner, 2000.

[BNSNT$^+$95] J. Bazan, H. Nguyen Son, T. Nguyen Trung, A. Skowron, and J. Stepaniuk. Application of modal logics and rough sets for classifying objects. In M. De Glas and Z. Pawlak, editors, *Proceedings of the Second World Conference on Fundamentals of Artificial Intelligence (WOCFAI'95)*, pages 15–26, Paris, 1995.

[BS98]   T. Bittner and J. G. Stell. A boundary-sensitive approach to qualitative location. *Annals of Mathematics and Artificial Intelligence*, 24:93–114, 1998.

[BS00]   T. Bittner and J. Stell. Rough sets in approximate spatial reasoning. In *Proceedings of the Second International Conference on Rough Sets and Current Trends in Computing (RSCTC'2000)*. Springer Verlag, 2000.

[CV95]   R. Casati and A. Varzi. The structure of spatial localization. *Philosophical Studies*, 82(2):205–239, 1995.

[EF91]   Max J. Egenhofer and Robert D. Franzosa. Point-set topological spatial relations. *International Journal of Geographical Information Systems*, 5(2):161–174, 1991.

[EFG97]   M. J. Egenhofer, D.M. Flewelling, and R.K. Goyal. Assessment of scene similarity. Technical report, University of Maine, Department of Spatial Information Science and Engineering, 1997.

[Gea66]   P. Geach. Some problems about time. *Proceedings of the British Academy*, 11, 1966.

[GFA93]   J. Glasgow, S. Fortier, and F.H. Allen. Molecular scene analysis: Crystal structure determination through imagery. In L. Hunter, editor, *Artificial Intelligence and Molecular Biology*. AAAI/MIT Press, 1993.

[LC97]   T.Y. Lin and N. Cercone, editors. *Rough Sets and Data Mining. Analysis of Imprecise Data*. Kluwer Academic Publishers, Boston, Dordrecht, 1997.

[Lin95]   T.Y. Lin, editor. *Proceedings of the Workshop on Rough Sets and Data Mining at 23rd Annual Computer Science Conference*, Nashville, Tenessee, 1995.

[NSR92]   R. Nowicki, Slowinski, and J. R., Stefanowski. Rough sets analysis of diagnostic capacity of vibroacoustic symptoms. *Journal of Computers and Mathematics with Applications*, 1992.

[Paw82]   Z. Pawlak. Rough sets. *Internat. J. Comput. Inform*, 11:341–356, 1982.

[RCC92]   D. A. Randell, Z. Cui, and A. G. Cohn. A spatial logic based on regions and connection. In *3rd Int. Conference on Knowledge Representation and Reasoning*. Boston, 1992.

[Sim87]   P. Simons. *Parts, A Study in Ontology*. Clarendon Press, Oxford, 1987.

[SV97]   B. Smith and A. Varzi. Fiat and bona fide boundaries: Towards an ontology of spatially extended objects. In S. Hirtle and A. Frank, editors, *Spatial Information TheoryA Theoretical Basis for GIS, International Conference COSIT '97, Laurel Highlands, PA*, volume 1329 of *Lecture Notes in Computer Science*, pages 103–119. Springer-Verlag, Berlin, 1997.

[Tho99]   Simon Thompson. *Haskell: The Craft of Functional Programming*. Addison-Wesley, 2 edition, 1999.

Join Indices as a Tool for Spatial Data Mining

Karine Zeitouni[1] , Laurent Yeh[1] , Marie-Aude Aufaure[2]

[1] Prism Laboratory - University of Versailles
45, avenue des Etats-Unis - F-78 035 Versailles Cedex
{last name.first name}@prism.uvsq.fr
[2] LISI Laboratory - INSA of LYON bat. 501
F-69 621 Villeurbanne Cedex
Marie-Aude.Aufaure@lisi.insa-lyon.fr

Abstract. The growing production of maps is generating huge volume of data stored in large spatial databases. This huge volume of data exceeds the human analysis capabilities. Spatial data mining methods, derived from data mining methods, allow the extraction of knowledge from these large spatial databases, taking into account the essential notion of spatial dependency. This paper focuses on this specificity of spatial data mining by showing the suitability of join indices to this context. It describes the join index structure and shows how it could be used as a tool for spatial data mining. Thus, this solution brings spatial criteria support to non-spatial information systems.

1. Introduction

In the area of Knowledge Discovery in Databases (KDD), Spatial Data Mining (**SDM**) is becoming an important issue that offers new prospects for many data analysis applications. Spatial data mining performs data driven analysis of large size spatial databases. The propensity of such large databases is due to the easiness of acquiring geodata. Indeed, geodata sets become available via Internet and the development of geo-coding tools allow attaching spatial data to a-spatial database according to address fields.

Spatial data mining tasks are considered as an extension of Data Mining (**DM**) tasks [11] in which spatial data and criteria are combined. As According to [15], these tasks aim at: (i) summarising data, (ii) finding classification rules, (iii) making clusters of similar objects, (iv) finding associations and dependencies to characterise data, and (v) looking for general trends and detecting deviation. They are carried out using specific methods, some of which are derived from statistics and others from the field of machine learning. As shown in [35], both approaches converge in the way they use spatial relationships. The problem is that the implementation of all those methods is not straightforward and right now, there is a lack of spatial data mining tools.

The goal of this paper is to introduce the use of the well-known "join index" structure as a tool for spatial data mining. Indeed, we point out that one major difference between traditional data mining and spatial data mining is the notion of spatial relationships. These spatial relationships correspond to the spatial join operator

J.F. Roddick and K. Hornsby (Eds.): TSDM 2000, LNAI 2007, pp. 105-116, 2001.
© Springer-Verlag Berlin Heidelberg 2001

and could be resolved by using "join index". We outline that this join index is a powerful tool to transform traditional database queries into spatial ones. A basic relational table is thereby sufficient to represent this join index. This has the great advantage to reduce spatial data mining to conventional data mining and provides an easy and efficient support of spatial data mining.

The rest of the paper is organized as follows: In the next section, we briefly introduce related works in the area of spatial data mining. In section 3 we emphasize the particularity and the importance of spatial relationships. These spatial relationships are then materialized by join indices in section 4. In this section, we outline the similarities between "join index" and "contiguity matrix" used in the field of spatial statistics. Section 5 discusses technical aspects in their implementation. We conclude in the last section.

2. Spatial Databases and Data Mining

This section briefly introduces spatial databases, and then related works in the fields of spatial statistics and spatial databases are described.

2.1. Spatial Databases Context

Spatial database systems are a great part of Geographical Information Systems (GIS) [22, 21, 23]. They store and manage huge volume of geographical entities such as road sections or lakes. Each entity combines the location description and other a-spatial data related to the entity such as the lake name or capacity.

A spatial database is organized in a set of thematic layers. A thematic layer is a collection of geographical objects that share the same structure and properties. A theme can represent a road network, and another can represent towns. This allows to selectively use the relevant themes for a specific purpose.

2.2. Advances in Spatial Data Mining

The aim of Spatial Data Mining [29] is to extract knowledge, spatial interactions and other properties that are not explicitly stored in the database. This process inherits from traditional data mining which could be performed by "a set of tools that allow to extract automatically or semi-automatically interesting and understandable knowledge (rule, regularities, patterns, associations..) from a database" [11].

The specificity of SDM lies in its interaction in space. In effect, a geographical database constitutes a spatio-temporal continuum in which properties concerning a particular place are generally linked and explained in terms of the properties of its neighbourhood. We can thus see the great importance of spatial relationships in the analysis process. Temporal aspects for spatial data are also a central point but are rarely taken into account.

Since traditional data mining methods [11] do not support location data or the implicit relationships between objects, it is necessary to develop specific methods for spatial data mining. As it is well known, geometric data and processing are more complex than traditional ones. Spatial applications also generate a huge volume of data. For these reasons, calculating these spatial relationships is time consuming. One major problem is to optimise analysis methods taking into account the huge volume of data and the complexity of spatial relationships.

Research in the field of SDM comes from both spatial statistics and databases. There is an abundant bibliography about spatial analysis [12, 13] existing before the GIS era. Many research works have been done to measure global and local spatial auto-correlation [7]. In the field of geostatistics [26], *kriging* technique analyses spatial trends. Recent research in interactive techniques for Exploratory Spatial Data Analysis (ESDA) have been developed [2, 1, 19, 20]. Openshaw [28] has developed a prototype using parallel computing to identify clusters. Some research works have also been done to extend the multivariate statistical analysis in order to support contiguity constraints [5]. From our point of view, spatial statistics methods are part of spatial data mining, since they provide data-driven analysis. Some of those methods are now implemented in operational GIS or analysis tools.

In the field of databases, spatial data mining algorithms have been proposed and prototypes have been developed. GeoMiner [14, 15] is an extension of DBMiner that integrates OLAP techniques and a coupling between a GIS and a database. Most of the proposed algorithms in GeoMiner are based on a priori knowledge: the concept hierarchies. Ester at al. devised a structure-of-neighbourhood graph[10], on which some algorithms are based. They have also worked on SDM methods such as clustering method (extension of DBSCAN with an R*Tree), classification (extension of ID3 and DBLearn), association rules (based upon an efficient spatial join), characterisation and spatial trends [10]. STING (University of California) uses a hierarchical grid to perform optimisation on the clustering algorithm [33]. We might also mention work on data warehousing dedicated to spatial data (University of Laval) [3].

In conclusion to this brief state of the art, we want to point out the similarities of the spatial statistics approach and the database one. The main similarity is the use and the importance of neighbourhood relationships. Contiguity matrices are used to compute spatial auto-correlation and neighbourhood graphs represent a secondary structure useful for many SDM algorithms. Another similarity is the current use of distance criteria in defining neighbouring objects as well as in clustering methods.

3. Spatial Relationships

Spatial relationships represent an essential characteristic in real world. They show spatial influences between entities. Indeed, observations located near to one another in space tend to share similar attribute values. This is known as one of the "1st law in geography" specified by Tobler [31]. Moran has defined a measurement of spatial auto-correlation between nearby data since 1948 [7]. Anselin has refined this in local auto-correlation indices [2] qualifying the correlation of each entity value with the values of its neighborhood.

Spatial and local auto-correlation only consider the interactions within one theme. In reality, thematic layers are often strongly correlated. For instance, precipitation and population density maps are correlated: the population depends on the agricultural production that depends on the precipitation. For this reason, we distinguish two kinds of relationships. On the one hand, those that link objects of the same thematic layers (we call it intra-theme). On the other hand, those involving two different layers (called inter-themes), like the inclusion of an accident location within a road section or within a county.

Spatial relationships have also been formalized in spatial database theory and extended to many topological relationships [8] such as intersection and inclusion. Moreover, metric relationships, e.g. using distance criteria, are also admitted as spatial relationships.

Spatial data mining methods make intensive use of spatial relationships. That actually distinguishes these methods from conventional data mining methods. This shows the main role of such relationships in spatial data analysis and mining. However, this uncovers many problems and specific properties as explained below.

First, those relationships are usually implicit except in topological models [4]. Their resolution leads to spatial join that is known as a complex and costly operation in spatial database. Therefore, efficient support of spatial joins should be studied.

Second issue is to allow specifying those relationships as spatial knowledge and to use it in the analysis. Indeed, some spatial relationships are pre-established and constitute spatial integrity constraints. For instance, car crash is constrained to be located on a road section. Hence, methods such as clustering should be modified to consider a linear network instead of an open space.

Third, inclusion spatial relationships, and more generally, the difference in the sizes or the weights of spatial units [23, 16], make it difficult to assess quantitative data. This involves the extension of those relationships to weighted relationships or more generally using a complex model including behavioral functions.

The last two points require the spatial relationships resolution, which is precisely the first point. This point is mainly the focus of our current work.

4.   Spatial Join Index

As contiguity matrix is an important component in spatial statistics, we show in this section that join index is also an important component for spatial data mining.

First, we describe the idea of a join index, then, discuss its different extensions to spatial databases. Section 4.3 shows its similarity with contiguity matrix allowing its direct utilisation in spatial statistics, such as auto-correlation computation, as well as for spatial data mining. Section 4.4 exposes another advantage of spatial join index, using it to simplify spatial analysis by reducing it to relational case. This is possible thanks to the very simple representation of a spatial join index by a relational table.

4.1. Join Indices

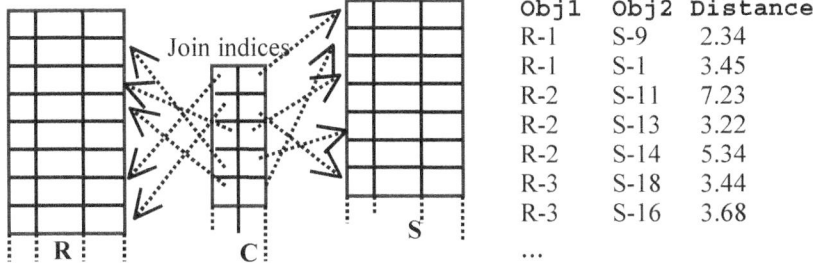

Obj1	Obj2	Distance
R-1	S-9	2.34
R-1	S-1	3.45
R-2	S-11	7.23
R-2	S-13	3.22
R-2	S-14	5.34
R-3	S-18	3.44
R-3	S-16	3.68
...		

Figure 1: Join index.   **Figure 2**: Spatial filter join index

The join-index method has been proposed by [32] as a valuable technique for increasing speed of query evaluation in Data Base Management System (DBMS) [23]. It constitutes a table that stores indice pairs (called join indices), each referencing an object of each input object collection (c.f. figure 2). Figure 1 shows this structure where C is the join index an **R** and **S** the two object collections. Here **R-1** to **R-n** are object identifiers [18]. The pairs of indices refer to objects that match the join criterion. In this figure, the join indices represent the distance between two objects (representing entities). In spatial topological relationships like inclusion, intersection, the third column could disappear because the distance is zero and does not change.

4.2. Extensions to Spatial Join Indices

Conversely to relational joins, spatial joins use as many criteria as possible spatial relationships. There are at least three alternatives to extend join indices to spatial databases.

- The first is a direct application, leading to build one join index set for each spatial predicate. That means exact computation of all these predicates, which is time and storage space consuming.

- A second way is to build only one join index set by adding columns for each kind of spatial relationship as in [10]. This improves a little bit the storage volume but the computing time is still high.

  The two previous solutions are not sufficient to deal with distance based joins when the distance is specified dynamically.

- The third solution proposes to store a coarse computation of the spatial criteria rather than the exact spatial join criteria. This is allowed by the definition of one join index having a third column for a distance value of the referenced spatial objects [30]. This structure (see figure 2) will play the role of a filter for most spatial joins. Indeed, the topological relationships as well as the metric ones can be deduced from using a simple selection in this join index. Notice that in this solution as well as for any distance based criteria, one usual optimisation of storage space is to limit the distance to a given

Karine Zeitouni, Laurent Yeh, and Marie-Aude Aufaure

maximal distance (called scope). Indeed, instead of storing all combinations of couples of object references, only those having a reasonable distance (under the scope) will be retained. Another alternative for optimising distance based join indices is described in [34].

4.3. Homogeneity with Contiguity Matrix

Besides, in spatial analysis area, as for the computation of Moran's and Geary's indices [7], it is usual to use what is called "contiguity" matrix. This matrix is defined as M where $M(i,j) = 1$ when the two objects i and j are contiguous, and $M(i,j) = 0$ otherwise. As this is a sparse matrix, it is usual to represent it in a more compact format as an array structure. Actually, it is the same structure than the above join index by using as object identifiers the row numbers. This homogeneity means that the same structure could serve in the new spatial data mining methods and the statistical spatial analysis methods. The second interest is to provide a generic way to yield contiguity matrices under any criteria.

Notice that contiguity matrices are square matrices defined on a unique data set (i.e. intra-theme) while join indices could apply to compare two data sets (i.e. inter-themes). So, spatial join indices are more general than contiguity matrices.

4.4. Reducing to Relational DM

After the computation step of join indices, using them has three main advantages that we outline in this sub-section. The first advantage is to be able to provide spatial analysis functionalities to a system that do not have this capability. The second one is to allow the use of powerful query language. The third one is that such a structure is used as an accelerator.

For the first advantage, join indices could be implemented using an external process to the given system. With modern system capabilities as DLL (Dynamic Link Library) or IPC (Inter-Process Communications) and Internet capabilities (using Socket), it is easy for most widespread tools to raise data analysis in spatial domain. This means a way for spatial data mining. This extensibility is described with technical aspects in the next section.

For the second advantage, as a join index has a tabular representation in relational DBMS, it could be handled in the same way as other tables. This allows using all the power of database query languages such as the normalized SQL language. For instance, by selecting and matching entities according to their attributes, one can focus the analysis on relevant data.

For the third advantage, many optimising techniques in spatial database domain have aimed at increasing the performance of spatial data retrieval including spatial joins. One category uses spatial access methods (e.g. R-tree, Grid-file, etc.). Here, using join indices is another alternative, based on the principle of pre-computing join operations. The interest in pre-computing is to do it once and to use it many times allowing an important gain while producing the final join result. Indeed, for the analysis purpose, interactive hypotheses tests become possible.

Performance measures have shown the gain in terms of processing time by comparing execution time with and without the pre-computed index (see figure 3).

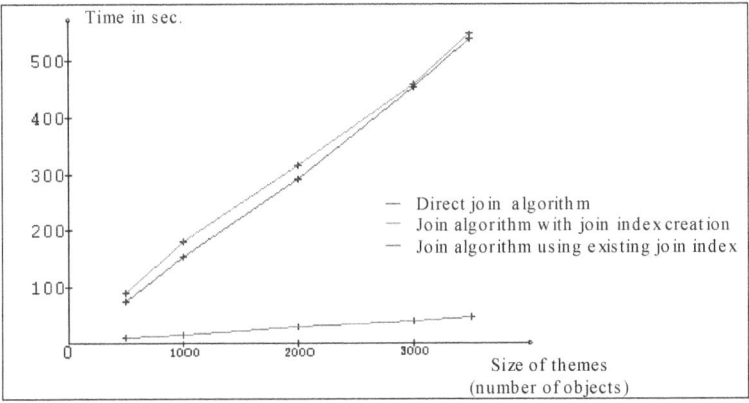

Figure 3: Performance measurement in using spatial join index

In many applications, the construction of join index structures could be done while loading data in the system. This happens in reality very few times. In addition, this computation could be done offline.

In summary, due to its tabular representation, join index is a pragmatic and efficient tool for spatial data mining.

4.5. Join Index Use in SDM

There exist two possibilities in using spatial join indices for SDM purpose:

1. during the stage of data selection preparing them to SDM tasks: as it is well known, previously to DM, KDD process includes a phase that transforms the initial data to a target data set. This data set holds relevant information for the analysis and should have the expected format of DM algorithms. The most used format is one table. This table is generally built by joining many initial database tables holding the relevant attributes. This corresponds to de-normalization in relational databases. Following this process, a pragmatic approach to spatial KDD is to, first, spatially join relevant thematic layers and then, apply DM tasks.

2. by strongly integrating them in specific SDM algorithms [9]. Here, join indices are viewed as a low-level library only manipulated internally. This is more complex to implement and is less portable to existing systems. However, strong integration achieves better performance in term of processing time.

5. Spatial Join Index Implementation

This section discusses technical aspects of spatial join index for a user who wants to implement them within an existing DBMS. It shows, first, the use of join indices at the user level. Then, it raises the problem of object referencing and studies diverse solutions. At the representation level, two possibilities are described and compared. Then an implementation approach is given.

5.1. How Does It Appear for Users ?

For users, join indices appear as normal tables of the database schema. So they can be manipulated using SQL. For instance, in the query of figure 4, where each of R and S contain a geometric attribute and the join index `Indices_dist_RS` stores all distances between R and S, this spatial join is replaced by two joins (figure 4, lines (1) and (2)). In this case, relation databases are sufficient, once join indices have been computed.

```
Select R.name, R.Population

From R, S, Indices_dist_RS I
Where I.obj1 = R.id          (1)
And    I.obj2 = S.id         (2)
And    I.distance <   50
```

Figure 4: Use of join indices in a query.

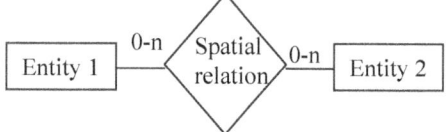

Figure 5: Spatial relation as a semantic link between entities.

This is a simple way to consider spatial relation evaluation. Indeed, this works as if all spatial relationships were semantic relations (in the well-known Entity/Relation data model [25]) and defined on each combination of spatial entities. Relation (diamond-shaped on figure 5) traduces a semantic relationship between two entities (rectangular in figure 5). This diagram shows the conceptual level. It is implemented in relational data model by tables. Thus, join indices tables are the physical representation of spatial relations. This simplifies the concepts for users by replacing complex spatial join concept by most familiar table joins.

Indeed, the main difference is that join indices are automatically deduced by a program instead of being a priori entered information. It is unrealistic to define conceptual model with all the above combinations of links, because it is a combinatorial problem. Only the useful combination could be pre-computed by the database designer.

5.2. Object Reference Choice

One main difficulty in using join indices, is to preserve a unique reference to a spatial value (similar to Object Identifier). Very few relational DBMS provide at the user's level an Object Identifier [18] even if in Object DBMS, there is the notion of Object

Identifier. In relational DBMS, the closest notion is the key attributes. However, the key could change by updating. Moreover, as a result of a query, we could lose the key attribute. Another possibility is to use the object row number in the table. This corresponds to the formal definition of contiguity matrices and systems like ArcView and Oracle allow handling rows. But, row numbers are no more available when objects are deleted or inserted. For these reasons, the use of join indices at the user's level in relational DBMS could be tedious. Integrity constraints or triggers could be implemented to prevent the key change. Fortunately, spatial databases are rather static.

5.3. One or Several Join Index Sets between Entities

Between two object sets, many spatial relationships (inclusion, neighborhood...) could be interesting. The problem is how to manage all these relationships in join indices [26]. Two approaches could be used to represent these data. The first approach is to store all the couples of indices that represent one relationship between two entities in individual sets. Then, we could name each set as the relationship. In the second approach (figure 6), we use a flag for the existence of a relationship between two entities. Thus, in one record, we could set several relationships. As illustrated in figure 6, the two first values are references to respective objects. The value '1' means the existence of a given relationship and null value for none relationship.

....
<s1, s2, null, null, 1, null, 1,...>
<s1, s3, 1, null, 1, null, null,...>
...

Figure 6: Example of indices with multiple relationships.

	Advantage	Disadvantage
Using several join indices (**SJI**) sets	Each set uses the requisite memory	Less efficient for multi-criteria queries
Using one join index (**OJI**) set	Fast to process	Many empty values (few objects match for a given relationship).

Figure 7: One set versus several sets.

Figure 7 compares the advantages and the disadvantages of each approach. Using several join indices (SJI) sets is optimal in the set size. Indeed, each set holds only a given spatial relationship. Each set stores only the couple of indices that match a given spatial relationship. However, when a query uses several spatial criteria (e.g. forest close to lakes AND included in Versailles City), it involves several join processing (one for each spatial relationship and one to compute the final result). Then, for SJI, the storage is more efficient but the processing cost could be important because of join operators and the query expression could be more complex.

On the contrary, using one join index (OJI) requires only two join operators for each pair of sets involving one or several (multi-criteria) spatial relationships within one query. However, the size of the set could be huge in case it integrates distance relationship. Indeed, distance involves all the couples of entities. The main consuming space is due to the two object identifiers. To overcome this size problem, a solution could be to represent in one bit a topological relationship. Moreover, we could represent an exclusive data between distance and a set of topological relationships. This needs one more bit to switch between these two representations.

5.4. Extensibility of Existing Tools

Join index provides a pragmatic solution for existing tools and DBMSs that do not handle spatial features or for GISs that do not support some spatial join criteria. For instance, in ArcView, distance based spatial joins are not supported. An external program could compute it and it appears as a simple table that could be integrated in the system.

We have applied this approach in our server prototype. The `Arcview` system sends a signal to server for computing a set of join indices. Then the result is stored as a `dbf` data file (directly readable in `Arcview`), and the server sends a signal to `ArcView` for synchronizing the next step of processing. This simple technique allows integrating in `ArcView` more complex processing. Then, using join indices could increase the extensibility of current data analysis system (as statistical tools).

6. Conclusion

In this paper, we emphasize that spatial relationships are of great importance in spatial data analysis and are used intensively in spatial data mining. This notion of spatial relationships does make the difference between data mining on alphanumerical data and on spatial data. It is represented by different concepts, such as contiguity matrix or neighborhood graphs, used in spatial analysis and spatial data mining.

In this paper, spatial relationships are materialized by an extension of the well-known join indices. In addition to its traditional use as an accelerator in the process of spatial join, this concept is deviated here in order to specify it as an efficient tool for spatial data mining. The definition of this tool is the main contribution of this paper. This tool is very simple to use thanks to its representation using the relational paradigm. Join indices can be handled in the same way than other tables and manipulated with the powerful and standardized SQL query language. We have also proposed a software architecture schema allowing the integration of such a structure into an existing system. Join indices could be handled as an external process of any system that do not provide any spatial functionality.

We have shown the great interest of the compatibility of this structure with contiguity matrices. Join indices can be viewed as an implementation of contiguity matrices allowing analysis on large sets of data. Once join indices are pre-computed, they may be used to apply analysis methods such as spatial auto-correlation, spatial characterization or spatial decision trees [10].

This paper also argues about the possibilities for coding join indices. A prototype have already be developed using the ArcView environment according to a SJI coding choice described in this paper. A second prototype is under construction with respect to the software architecture described in this paper. A short-term perspective is to compare in terms of performance the two different coding possibilities, i.e. OJI and SJI.

In general context of spatial data mining, there is many research issues. One issue concerns user assistance and user involvement in the process of knowledge discovery. Indeed, results interpretation could be sometimes a very hard task. Moreover, users should be able to intervene during the process to orient or to filter the analysis. The application of techniques coming from the field of visualization and interaction is already proposed in this objective [6, 17] in the context of conventional data mining. Extending these concepts to spatial data mining such as providing visualisation techniques for spatial relationships, including the user in the whole process of spatial data mining is an interesting perspective.

References

1. Andrienko, N. and Andrienko, G.: Interactive Maps for Visual Data Exploration, International Journal of Geographical Information Sciences 13 (4), pp. 355-374 (1999). See also URL: http://borneo.gmd.de/and/icavis.
2. Anselin, L.: Local indicators of spatial association - LISA. *Geographical Analysis*, 27, 2, pp. 93-115 (1995)
3. Bédard, Y., Lam, S., Proulx, M.J., Caron, P.Y. and Létourneau, F.: Data Warehousing for Spatial Data: Research Issues, Proceedings of the International Symposium Geomatics in the Era of Radarsat (GER'97), Ottawa (1997) pp. 25-30
4. Bennis K., David B., Quilio I., Thévenin J-M. and Viémont Y..: GéoGraph : A Topological Storage Model for Extensible GIS:, Proc. of Auto-Carto'10, Baltimore, USA, 368-392 (1991)
5. Burtschy, B. and Lebart, L.: Contiguity analysis and projection pursuit. In : *Applied Stochastic Models and Data Analysis*, R. Gutierrez and M.J.M. Valderrama, Eds, World Scientific, Singapore, pp. 117-128 (1991)
6. Card, S.K., Mackinlay, J.D. and Shneiderman, B.: Readings in Information Visualization: Using Vision to Think, Morgan Kaufmann (1999)
7. Cliff A.D., Ord J.K.,: Spatial autocorrelation, Pion, London (1973)
8. Egenhofer M.J. and Sharma J.: Topological Relations Between Regions in R2 and Z2, Advance in Spatial Databases, 5th International Symposium SSD'93. pp 316-331. Singapore (1993) Springer-Verlag.
9. Ester, M., Frommelt, A., Kriegel, H.-P., Sander, J.: Algorithms for Characterization and Trend Detection in Spatial Databases", *Proc. 4th Int. Conf. on Knowledge Discovery and Data Mining,* New York, NY (1998).
10. Ester, M., Kriegel, H.-P., Sander, J.: Spatial Data Mining: A Database Approach, Proceedings of the 5th Symposium on Spatial Databases, Berlin, Germany (1997)
11. Fayyad et al.: Advances in Knowledge Discovery and Data Mining, AAAI Press / MIT Press (1996)
12. Fisher, M. and Getis, A.: spatial analysis – spatial statistics, behavioural modelling and neurocomputing, Berlin, Springer (1997)
13. Fotheringham, S. and Rogerson, P.: Spatial Analysis and GIS, Taylor and Francis (1995)

14. Han J., Cai Y. and Cerone N.: Knowledge Discovery in Databases; An Attribute-Oriented Approach., Proceedings of the 18th VLDB Conference. Vancouver, B.C (1992) pp. 547-559. See also URL : http://www.cs.sfu.ca/~han

15. Han J., Koperski K., and Stefanovic N.: GeoMiner: A System Prototype for Spatial Data Mining, Proc. 1997 ACM-SIGMOD Int'l Conf. on Management of Data (SIGMOD'97), Tucson, Arizona, May1997 (System prototype demonstration).

16. Holt, D., Steel D.G., Tramer M.: Area Homogeneity and the Modifiable Areal Unit Problem, *Geographical Systems (3)*, pp. 181-200 (1996)

17. Keim, D.A., Kriegel, H.P.: Visualization Techniques for Mining Large Databases: A Comparison, IEEE Transactions on Knowledge and Data Engineering, vol 8, n°6 (1996)

18. Khoshafian, S.N, and Copeland, G.P: Object Identity. In Proc. of the ACM Conf. on Object-Oriented Programing Systems and Languages (OOPSLA), pages 408-416. (1986)

19. Kraak, M.J. and MacEachren, A.M.: Visualisation for exploration of spatial data. International Journal of Geographical Information Sciences 13 (4), pp. 285-287 (1999)

20. Kraak, M.J.: Visualizing spatial distributions. Chapter 11 in Longley, P., M. Goodchild, D. Maguire & D. Rhind (editors) Geographical information systems: principles, techniques, management and applications. New York: J. Wiley & Sons (1999) pp.157-173.

21. Laurini R., Thompson D.: Fundamentals of Spatial Information Systems, Academic Press, London, UK, 680 p, 3rd printing (1994)

22. Laurini, R.: Information Systems for Urban Planning : A Hypermedia Cooperative Approach , Taylor and Francis (2000)

23. Longley, P.A., Goodchild, M.F., Maguire, D.J., Rhind, D.W. : Geographic Information Systems , Volume 1, Wiley, 1999

24. Lu, W. and Han, J: Distance-Associated Join Indices for Spatial Range Search. Eighth International Conference on Data Engineering, (1992) Tempe, Arizona, pp. 284-292

25. Maier, D., The Theory of Relational Databases, Computer Science Press, 1983.

26. Matheron, G. : Principles of geostatistics. *Economic Geology*, 58, pp. 1246-1266, (1963)

27. O' Neil, P. and Graefe, G: Multi-tables joins through bitmapped join indices. SIGMOD Record, 24(3), pp. 8-11 (1995)

28. Openshaw, S., Charlton, M., Wyme,r C. and Craft, A: A mark 1 geographical analysismachine for the automated analysis of point data sets, International Journal of Geographical Information Systems, Vol. 1 (4), pp. 335-358 (1987). See also URL : http://www/ccg.leeds.ac.uk/smart/gam/gam.html

29. Roddick, J.F, Spiliopoulou, M.: A Bibliography of Temporal, Spatial and Spatio-Temporal Data Mining Research, ACM SIGKDD Explorations, volume 1, Issue 1 (1999)

30. Rotem D: Spatial join indices, Proc. of 7th Conf. on Data Engineering, Kobe, Japan (1991) pp. 500-509

31. Tobler W. R.: Cellular geography, In Gale S. Olsson G. (eds.) Philosophy in Geography, Dortrecht, Reidel (1979) 379-386

32. Valduriez P., "Join indices", *ACM Trans. on Database Systems*, 12(2); 218-246, June 1987.

33. Wang, W., Yang, J. and Muntz, R.: STING+: An approach to active spatial data mining, Proceedings of the Fifteenth International Conference on Data Engineering, Sydney, Australia. (1999) IEEE Computer Society. 116-12

34. Yeh, T-S: Spot: Distance based join indices for spatial data, ACM GIS 99, Kansas City, USA, pp 103-110 (1999)

35. Zeitouni K.: A Survey on Spatial Data Mining Methods Databases and Statistics Point of Views, Information Resources Management Association International Conference (IRMA'2000), Data Warehousing and Mining Track, Anchorage, Alaska, USA (2000)

Data Mining with Calendar Attributes

Howard J. Hamilton and Dee Jay Randall

Department of Computer Science, University of Regina
Regina, Saskatchewan, Canada, S4S 0A2
{hamilton,randal}@cs.uregina.ca

Abstract. This paper addresses the problem of data mining from temporal data based on calendar (date and time) attributes. The proposed methods uses a probabilistic domain generalization graph, i.e., a graph defining a partial order that represents a set of generalization relations for an attribute, with an associated probability distribution for the values in the domain represented by each of its nodes. We specify the components of a domain generalization graph suited to calendar attributes and define granularity, subset, lookup, and algorithmic methods for specifying generalizations between calendar domains. We provide a means of specifying distributions. We show how the calendar DGG can be applied to a data mining problem to produce a list of summaries ranked according to an interest measure given assumed probability distributions.

1. Introduction

Temporal and spatial attributes are common in relational and object-oriented databases. A ***temporal attribute*** might be used to specify the beginning of an event, such as a *birth date* to mark the start of a new life, or the ending of an event, such as the *check out time* from a hotel, or the duration of an event, such as the *elapsed time* of a race. A ***spatial attribute*** might be used to specify a geographical location or a coordinate of an object in an arbitrary space.

Probabilistic domain generalization graphs can be used to guide the data mining process in the presence of domain knowledge about temporal and spatial attributes. Informally, a ***domain generalization graph*** can be thought of as a graph showing possible generalizations as paths through a graph. A formal definition is given in [9]. Each node in the graph corresponds to a domain of values. Each link in the graph corresponds to a generalization relation, which is an mapping from the values in the domain of the initial node to the final node of the link. Every domain generalization graph has a single source node, called the ***bottom***, corresponding to the most specific domain of values, which are typically the base data values present in the database. Every domain generalization graph also has a single sink node, called the ***top***, corresponding to a domain called *Any* with only one value. The bottom node is the ancestor of all nodes in the domain generalization graph, so unlike in trees, the child nodes are above the parent nodes.

Domain generalization graphs are appropriate for summarization data mining, in which interesting summaries of all or part of the data are sought [9]. The act of

J.F. Roddick and K. Hornsby (Eds.): TSDM 2000, LNAI 2007, pp. 117-132, 2001.
© Springer-Verlag Berlin Heidelberg 2001

creating a summary is a form of generalization.   Multiple possible paths of generalization are conveniently represented and manipulated using a domain generalization graph, because the set of paths through the graph corresponds to the set of possible generalizations that can be defined using the generalization relations. Wherever values at different levels of granularity are related by generalization relationships, domain generalization graphs should be considered as a possible representation.

In a probabilistic domain generalization graph, a description of a probability distribution is associated with each node in the graph to describe the expected distribution of the values in the domain. For example, if the domain of values is the names of the countries of the world and expectations are based on population, the distribution could be specified by giving each country's name associated with the ratio of that country's population to the world population.

A *calendar attribute* is one whose domain contains date and time values, such as birth dates and check out times, while a *duration attribute* is one containing durations [14]. A noteworthy problem in developing techniques for generalizing and presenting temporal data is that time can be represented in many ways depending on the context. Temporal values can be generalized in different ways and to different levels of granularity [11]. For example, if the calendar attribute in the data specifies login times, its values can be generalized to give the number of logins in (1) each hour (of each day) or (2) each of the seven possible days of the week. Choosing different levels of granularity for the first way can give values generalized to the day, the month, or the year. We present a revised domain generalization graph for calendar attributes by explicitly identifying the domains appropriate to the relevant levels of temporal granularity and the mappings between the values in these domains; an earlier version of this graph was presented in [15].

Here, we consider the problem of data mining in the presence of calendar attributes in relational databases using probabilistic domain generalization graphs. We use domain generalization graphs because they provide a graphical structure that can be used as both a navigational aid by the user and as a guide to heuristic data mining procedures. The problem addressed in this paper is as follows: given a probabilistic DGG for a calendar attribute and a relation containing values for that attribute, create a list of generalized relations (summaries) at various levels of generality ranked according to a specified numerical interest measure. Although only the highest ranked summaries are displayed, our method creates all distinct, nontrivial summaries that can be prepared. To create a summary, we perform generalization by transforming values in one domain to another, according to directed arcs in the domain generalization graph.

The remainder of this paper is organized as follows. In the following section, we review domain generalization graphs and present a particular graph for calendar attributes. We also describe four methods for generalizing temporal data for a calendar attribute. In Section 3, we relate our work to other recent research. In Section 4, we describe probabilistic domain generalization graphs and explain how probability distributions are attached to the nodes in such graphs. In Section 5, we describe an application of our methodology to a sample data set. Finally, in Section 6, we present our conclusions.

2. Generalizing Calendar Values

A *concept hierarchy* associated with an attribute in a database is represented as a tree, where leaf nodes correspond to the actual data values in the database, intermediate nodes correspond to more general representations of the data values, and the root node corresponds to the most general representation of the data values. Knowledge about higher-level concepts can be discovered through a series of transformations of specific values to more general values. search beginning at the leaf nodes using a process called *attribute-oriented generalization* [3,8]. If several concept hierarchies are associated with the same attribute, meaningful knowledge about the attribute can be expressed in different ways. Common attribute-oriented generalization methods require the user to select one concept hierarchy, ignoring the relative merits of other possible generalizations that could produce interesting results. To facilitate other possible generalizations, *domain generalization graphs* (DGGs) have recently been proposed [7,9]. A concept hierarchy is represented in a domain generalization graph as a sequence of nodes, with one node for each level of the concept hierarchy.

We now describe our calendar DGG for calendar attributes, a part of which is shown in Figure 1. This DGG is a refined version of one presented in [14]. In Figure 1, the node labelled *YYYYMMDDhhmmss* represents the most specific domain considered, i.e., the finest granularity of our calendar domain is one second. Higher-level nodes represent generalizations of this domain. The arcs between nodes represent generalization relations. To handle data with calendar values specified to finer granularity, e.g., microseconds, more specific nodes could be added to the DGG.

The calendar DGG can be used to guide the generalization of calendar data into higher-level concepts. In knowledge discovery, each traversal of an arc in the DGG corresponds to generalizing a set of data from the domain indicated by the initial node of the arc to the domain indicated by the terminal node of the arc. For example, we generalize from *YYYYMMDDhhmmss* to *YYYYMMDDhhmm* by removing the second information from the calendar attribute. Among the more complicated are the contrast between calendar months and lunar months. The arcs leaving the node labelled *YYYYMM* indicate that data can be generalized to the higher-level concept of *MM* (calendar month). For the *MM* concept we generalize a set of 28, 29, 30, or 31 days into a single calendar month. A *lunar-month* generalizes a set of 29 or 30 days into one lunar month as required to closely follow the actual lunar cycle. While a calendar month always starts at the beginning of a month (that is, DD = 01), a lunar month does not. Thus a *lunar-month* must be generalized from the node labelled *YYYYMMDD*. When a new representation is required in the calendar domain, this DGG can be extended by adding new nodes and arcs and by defining new generalization relations associated with the arcs.

Four types of generalization relations are associated with the arcs in a calendar DGG: granularity, subset, lookup, and algorithmic. For *granularity generalization*, we assume that *YYYYMMDDhhmmss* can be represented as six subattributes (*YYYY, MM, DD, hh, mm, ss*). We generalize by suppressing subattributes of the *YYYYMMDDhhmmss* attribute from least significant (*ss*) to most (*YYYY*). For example, to generalize from *YYYYMMDDhhmmss* to *YYYYMMDD*, the *ss*, *mm*, and *hh* subattributes are discarded. All five domains this creates are shown in Figure 1.

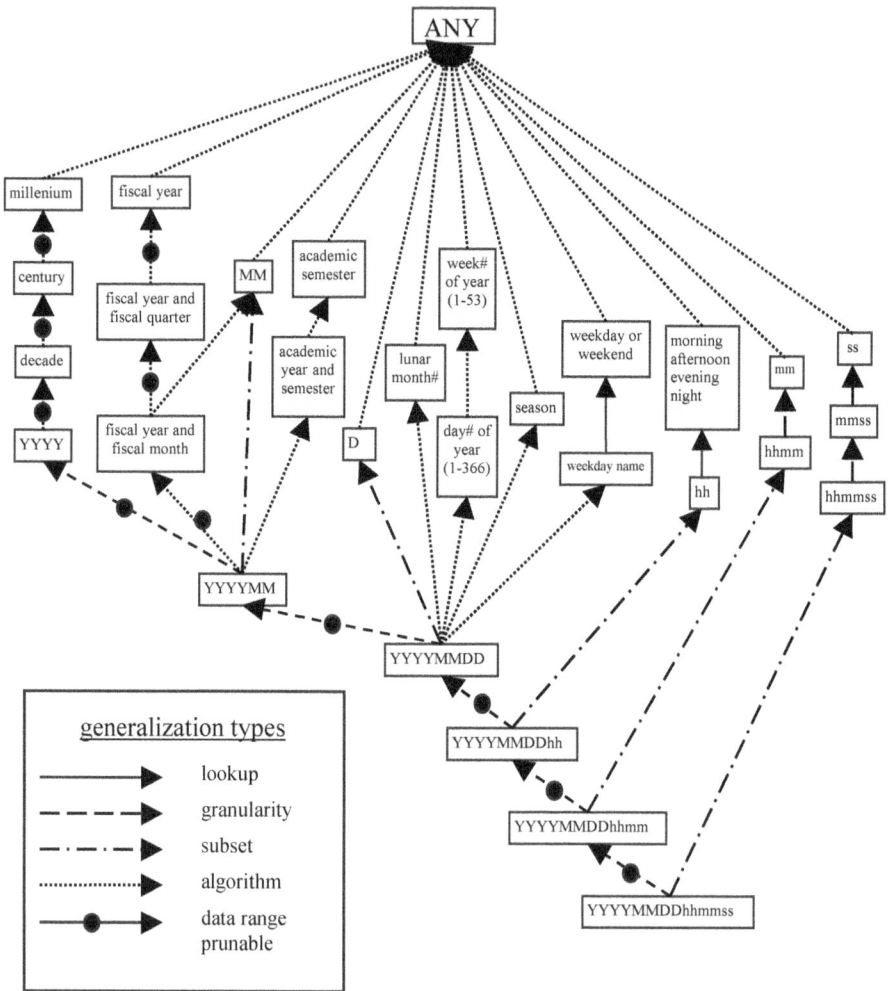

Figure 1. Domain Generalization Graph for a Calendar Attribute

In *subset generalization*, granularity generalization is extended so that any combination of the subattributes in *YYYYMMDDhhmmss* can be discarded. To generalize *YYYYMMDDhhmmss* to the periodic intervals of node *hh* (which could be used to discover that something happens every evening sometime in the hour starting at 9:00 pm), the *YYYY, MM, DD, mm,* and *ss* subattributes are discarded. The remaining subattributes do not need to be adjacent. This flexibility allows us to generalize to a node such as *MMhh*, which causes values to be categorized by both the month and hour. For simplicity, only a few of the domains that can be created by subset generalization are shown in Figure 1. Because subset generalization subsumes granularity generalization, the latter is treated as a special case of the former in our implementation. Nonetheless, we find it a useful distinction at the conceptual level.

In *lookup generalization*, the domain of the higher-level concepts is specified in a table. To generalize the calendar attribute with this method, the lower-level concept is looked up in the table to obtain the corresponding higher-level concept. For example, to define a generalization from *day of week* to *weekday or weekend*, we list all weekday values along with their generalizations in a table. To generalize a value we search the table for the day of the week and select the corresponding weekday or weekend value.

In *algorithmic generalization*, an algorithm is used to generalize a value from one level of generality to another. Although algorithm generalization subsumes the three previous types, we find the distinction useful and only refer to algorithm generalization when none of the others is applicable. For example, to determine whether a particular year is a leap year, or to determine the day of the week on which someone was born, a lookup table would be extremely large, while algorithm generalization allows convenient expression.

Additionally, it can be valuable to duplicate the calendar attribute in a relation so that it can be generalized to multiple nodes in the DGG simultaneously. An example of this is comparing the number of instances of an event per hour on weekdays versus weekends. The calendar attribute is duplicated and one of the copies is generalized to *weekday or weekend* while the other is generalized to *hh*. This technique reduces the need to define new complex nodes in the DGG. It is never valuable to duplicate an attribute and generalize one to the point of being an ancestor of the other.

3. Related Work

Most work related to time in Artificial Intelligence (AI) has concentrated on defining a time domain and facilitating reasoning about events. Time can be represented using timepoints, time intervals, or a combination of the two, and it can be linear, non-linear or branching. In relational databases, time is represented by database timestamps that include time and date and that correspond to global (or absolute) timepoints in many temporal representation schemes. Thus, for knowledge discovery in relational databases that include timestamps, we use timepoints to define our calendar DGG and the intervals that correspond to less granular time periods. For our purposes, time is defined as linear, discrete, and bounded.

Terenziani discusses how to compose and intersect temporal constraints, and in doing so combines temporal logic with user-defined calendars [16]. Euzenat applies the idea of spatial granularity to the time domain and uses temporal algebra to construct upward and downward granularity change operators for converting time relationships from one granularity to another [6]. These ideas form the basis for the granularity generalization technique given in Section 2.

Cukierman and Delgrande [5] construct a calendar structure and calendar expressions that form the basis for our definition of a calendar DGG. They describe a decomposition technique that is the inverse of the generalization used in a calendar DGG. In addition, their directed acyclic graph of one calendar structure corresponds to the inverse of one concept hierarchy (and thus one path in the calendar DGG). However, a DGG allows us to combine multiple calendar structures in the same representation, enabling us to take advantage of the duplication that exists between similar calendars. For example, a year is composed of 12 months, regardless of whether we are looking at a standard (Gregorian) year, an academic year, or a fiscal year.

Cukierman and Delgrande also define two orthogonal properties, *alignment* and *constancy*. Four types of decomposition for reasoning about schedulable, repeated activities follow from these two terms. *Aligned decomposition* means that the union of the submembers completely covers the original member (e.g., days from months). *Non-aligned decomposition* means that the union of the submembers does not completely cover the member leaving gaps at the beginning and/or end (e.g., weeks from months). *Constant decomposition* means that every time the member is broken into submembers, a constant ratio is maintained (e.g., days from weeks). A *non-constant decomposition* has a variable number of submembers (e.g., days from months).

For generalization, we define *compositions*, which are the inverses of Cukierman and Delgrande's decompositions. Four types of compositions can be defined analogously as *aligned composition, non-aligned composition, constant composition,* and *non-constant composition*. We intentionally do not include any non-aligned compositions in our calendar DGG. Instead, alternate paths of generalization from a common ancestor are explicitly shown. For example, months are not composed directly from weeks, but both months and weeks are composed from days in the calendar DGG. Constant composition is most appropriately implemented using a short algorithm, while non-constant composition is most appropriately implemented using a lookup table.

Other recent work has focussed on the connection between multiple temporal granularities and data mining. Granularity factors that affect data mining were described by Andrusiewicz and Orlowska [1]. Bettini et al. provide conventions similar to those given here for naming the multiple temporal granularities, although they do not provide a data structure similar to domain generalization graphs for representing the relationships between these granularities [2]. They also give preliminary ideas for applying their system in the context of data mining. Merlo et al. build on the work of Bettini et al. to specify the syntax and semantics of expressions involving data with multiple temporal granularities [12]. Combi et al. use a much simpler structure than our calendar DGG with an ordered granularity from SUP (top) to year, month, day, hour, minute, second, INF (bottom) [4]. However, they cannot handle phenomenon such as weeks.. Rainsford and Roddick provide a method for adding temporal semantics to association rules, based on a structured relationships among temporal relations, rather than our structure among domains [13]. Another approach to intensional query answering at multiple levels of abstraction is given by Suk et al. [17].

4.   Probabilistic Domain Generalization Graphs

Given a set of generalized relations, an interest measure can assign a numeric score to each. These scores can be used to rank the results. These scores can also be used to determine which generalized relations are consistent with an expectation and which ones conflict with it. Interest measures are computed by comparing an observed distribution to an expected distribution. The expectation of a tuple is the product of the expectations of the value in each of its component attributes.

The simplest approach is to assume a uniform distribution, but this approach requires specific assumptions about where the distribution is uniform. For example, at the *WeekdayOrWeekend* node (hereafter called *WDWE*), the domain consists of two elements and a uniform distribution gives an expected frequency of 0.50 for each of Weekend and Weekday, the two elements of the domain. Similarly, a uniform distribution among the seven elements of the *WeekdayName* node's domain, gives 0.14 for each of Sunday, Monday, etc. Unfortunately, these two distributions are inconsistent because only two days (Saturday and Sunday), with a total expectation of 0.29, are generalized to Weekend with an expectation of 0.50, while the remaining five days, with a total expectation of 0.71, are generalized to Weekday. To avoid such inconsistencies and generally simplify the process of specifying expectations, knowledge about expectations can be propagated among the nodes of the DGG. For example, more reasonable expectations of 0.29 for Weekend and 0.71 for Weekday result from propagating the distribution upward from *WeekdayName* to *WDWE*.

We define ***upward propagation*** as the process of translating a distribution from a node in the DGG up to at least one its child node(s) and possibly higher as well. The distribution at the higher level is the original distribution proportionately weighted according to the relevant generalization relation. Either a uniform or a non-uniform distribution can be upward propagated, but the distribution is assumed to be uniform unless otherwise stated. ***Bottom-up propagation*** propagates a distribution from the bottom node of the DGG to all other nodes of the DGG. ***Downward propagation*** is the process of propagating a distribution from a node to at least one of its parent nodes and possibly lower as well. ***Top-down propagation*** propagates a distribution from the top node of the DGG to other nodes of the DGG. Top-down propagation is based on the assumptions of uniformity at the most general level.

The concept hierarchies for the (WeekdayName, WeekdayOrWeekend, Any) path of the calendar DGG, as shown in Figure 2 illustrate upward and downward propagation. These concept hierarchies have the same nodes, but a different probability distribution is created based on the direction of propagation. For top-down propagation, consider the node *Any*, which is the top of the hierarchy of values shonw in Figure 2(a). The *Any* node has a domain of only one value, but this is assumed to represent a distribution, i.e., an expectation of 1 for this value. Since the single value at the top is created by generalizing from the two values in the domain of the *WDWE* node, we assume the distribution is 0.50 each. The distribution at the WeekdayName level follows from this assumption and further downward propagation.

As an example of upward propagation, we might assume that logins are uniformly distributed among the seven members of the *WeekdayName*, and propagate this distribution upward to obtain a 0.29 / 0.71 distribution for *WDWE*. In bottom-up propagation, *WeekdayName*'s distribution would itself be propagated upward from its parent nodes.

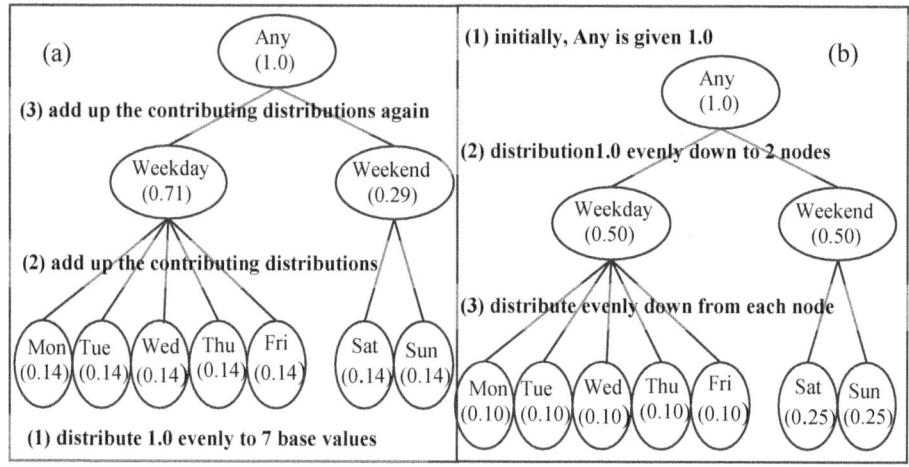

Figure 2. Propagation of Expectation Values: (a) upward, (b) downward.

The bottom-up approach always creates consistent distributions among all nodes in the DGG because the distribution at each node is consistent with the distribution of the base values at the bottom node. However, the top-down approach can suggest inconsistent distributions for a particular node in the DGG based on multiple paths down to it. Propagating a distribution from a single child down to a single parent gives an unambiguous result. Similarly, propagating from a single child down to multiple parents is equally straightforward. But when values are propagated down from multiple children to a single parent, it can be impossible to calculate consistent values without recalculating as much as the whole graph.

Four methods are provided for a user to specify expected distributions. The generalized relations are compared to these distributions. The methods are as follows. **(1) Data driven**: As the generalization space is generated, the algorithm records how the values for each attribute are generalized. When the generalization space is complete, the distribution for every node in every DGG is available. This method is based on bottom-up propagation of a *non-uniform* distribution. Its two principal weaknesses are: (1) if any base value does not occur in the data, it is assigned 0 in the distribution, and (2) since the expectation is derived purely from the data, there will be no differences between expectation and observation.

(2) Data dictionary: If the expected distribution of values for an attribute is known at for any node in the DGG, the user can specify this distribution in a table in the database. The table contains one column with all unique values for the domain and a second column with the corresponding expected frequency (in the range 0-1). An entry is added to the *DGG* file to specify the database table and column names. The expected frequencies in the data dictionary are propagated upward to all DGG nodes reachable from the specified node. They can also be propagated downward to parent, and then grandparent, etc. as long as the ancestor node has only one child node.

```
username date          in   -  out    duration
user328  Sun Jan 18 00:26 - 00:26   (00:00)
user328  Sun Jan 18 00:55 - 00:59   (00:03)
user645  Sun Jan 18 01:21 - 01:42   (00:20)
...
user602  Sat Jan 24 23:48 - 23:48   (00:00)
```

Figure 3. Sample 'last' output

(3) Explicit histogram: The finest degree of control is available with an explicit histogram specification. It can optionally be specified for each DGG node.   An example specification is as follows.

```
BeginExplicitHistogram
DGGNode=WeekdayOrWeekend
Weekday    0.714285714
Weekend    0.285714286
EndExplicitHistogram
```

The DGGNode line identifies the node. Each subsequent line contains a value occurring in that node's domain and the corresponding expected frequency. A warning is displayed if the sum of the expected frequencies exceeds one.   In the current implementation, no propagation occurs.

(4) Explicit uniform distribution: The distribution for all elements at some node in the DGG is specified as uniform with an optional `Explicit Uniform Distribution` specification. An example specification is as follows.

```
BeginExplicitUniformDistributions
WeekdayName      0.142857142
DayNumberOfYear 0.002739726
EndExplicitUniformDistributions
```

Each content line specifies the name of a DGG node and the expected frequency to be distributed to each value in the domain. This type of specification is provided as a convenience for cases where a data dictionary or explicit histogram would be cumbersome. As with explicit histograms, a warning message is displayed if the sum of the expected frequencies exceeds one, and no propagation is done in the current implementation.

Our implementation prioritizes these alternatives in the following order: explicit histogram, explicit uniform distribution, data dictionary, and data driven. Thus, a value from an explicit histogram overrides a value from an explicit uniform distribution, and so forth. An output summary is produced as comma-separated values, which are readily displayed and processed with Microsoft Excel and other standard tools.

5. Example Application

We conducted a series of experiments using data from login statistics from 1998-2000. We studied the usage of two computers, Hercules and Mercury, but since they were similar, here we describe only the Hercules results.

The input data are 523253 lines of output from the Unix "`last`" program, which produces a single line for each user session that shows login and logout times and the duration of the session. The input data were collected over a period of somewhat more than two years, from June 1998 to June 2000. Sample output from `last` is shown in Figure 3. We focus on the login times, which can be easily converted to a format that can be mapped to the *YYYYMMDDhhmm* node in the DGG. Times are not recorded to seconds. The problem is to find interesting summaries of the data at various levels of granularity. We considered a simpler version of the same problem with far less data in [15].

We used variance as the measure of interest. We computed variance of the fraction of all logins for the following distributions:

USER	LOGINTIME	#Tuples	Coverage	Variance
Any	WeekdayOrWeekend	2	1	0.1889
Any	FiscalYear	3	1	0.0505
Group	Any	12	1	0.0341
Group	WeekdayOrWeekend	22	0.917	0.0129
Any	AcademicYear	3	1	0.0106
Any	YYYY	3	1	0.0106
Group	FiscalYear	29	0.806	0.0066
Group	AcademicYear	29	0.806	0.0052
Group	YYYY	29	0.806	0.0052
Any	AcademicYearAndSem	7	1	0.0045

Table 1. Top Ranked Nodes for Hercules data and (C, U)

Distribution 1. (C, U): We assumed a uniform distribution at all nodes in the username and calendar DGGs, based on the set of all occurring values. Thus, for 700 users, we expected 1/700 of the logins for each user, but for 12 user groups, we expected 1/12 of all logins to each group, regardless of the number of users in each group. A uniform distribution was assumed for years (among 1998, 1999, and 2000 for YYYY), for months in general (MM), for the 36 months (YYYYMM), for the 31 possible days of the month (DD), for the 7 days of the week (DayOfWeek), for the 2 parts of the week (WeekdayOrWeekend), etc.

Distribution 2. (C',U): We used the uniform distribution U for the username attribute and an adjusted distribution C' for the calendar DGG with two types of adjustments. First, the distribution for MM was adjusted for the varying number of days in the months (28 to 31) and the distribution for WeekdayOrWeekend was adjusted for the varying number of weekdays (5) and weekend days (2) in a week. The adjusted distributions were propagated downward. Secondly, because we knew that only parts of 1998 and 2000 were present, a further adjustment was made starting at the YYYYMMDD node. The distribution was propagated upward to all reachable nodes.

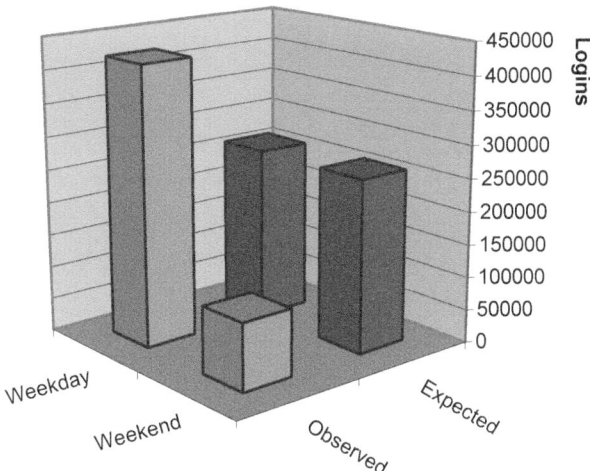

Figure 4. Observed and Expected Logins for (Any, WeekendOrWeekday)

After adjustment, the expected frequencies for the years were 0.25 for 1998, 0.60 for 1999, and 0.15 for 2000. The distribution at some nodes, such as YYYYMM, was affected by both upward and downward propagations. In these cases, we manually combined the distributions. For example, the information that in a typical year, 7 months have 31 days, 4 have only 30 days, approximately 0.25 have 29 days, and approximately 0.25 have 28 days was propagated downward. Suppose that the information that the data covered only the period 1 April 1998 to 30 April 2000 was propagated upward. Then the combined distribution would be based on the proportions: 14 with 31, 9 with 30, 1 with 29, and 1 with 28.

Distribtuion 3. (C', U'): We used the adjusted distribution C' for the calendar attribute and an adjusted distribution U' for the user attribute with somewhat arbitrary expectations based on our naïve beliefs about classes of users.

Based on these distributions, let us now examine a few of the highest ranked nodes in the generalization space and their corresponding graphs. As shown, in Table 1, for (C, U), the combination of the (any, WDWE) domains has the highest interest value. In other words, if the week was divided into two parts, weekdays and weekend, then the weekdays part had significantly more logins than the weekend part, as shown in Figure 4. This type of extremely obvious result is frequently obtained in data mining. With probabilistic DGGs, we have a convenient means of adjusting our expectations to avoid further repetition of both this obvious result and all its logical consequences.

As previously mentioned, the (C',U) distribution adjusts expectations to handle the inequalities in the number of days classified as weekdays (5) and weekend days (2), and all similar inequalities the calendar attribute. The interest ratings of nodes related to these concepts were lower with (C',U) than with (C, U). In the results shown in Table 2, (Group, Any) is ranked first instead of third as with (C, U). Also, (Any, WDWE) is second instead of first, and (Group, WDWE) is third instead of fourth.

USER	LOGINTIME	#Tuples	Coverage	Variance
Group	Any	12	1	0.0341
Any	WeekdayOrWeekend	2	1	0.0177
Group	WeekdayOrWeekend	22	0.917	0.0124
Group	FiscalYear	29	0.806	0.0063
Group	AcademicYear	29	0.806	0.0052
Group	YYYY	29	0.806	0.0052
Any	AcademicYearAndSem	7	1	0.0020
Any	AcademicYear	3	1	0.0015
Any	YYYY	3	1	0.0015
Any	FiscalYearAndQuarter	9	1	0.0013

Table 2. Top Ranked Nodes for Hercules data and (C', U)

The three nodes ranked highest in Table 2 are related. Investigation revealed that the (Group, Any) relation is unusual because csugrd and unknown have far more logins than expected while every other group has far fewer. This relationship is shown in Figure 5 for (Group, Any), where the X axis (across) gives the group, the Y axis (upward) gives the number of logins, the Z axis has observed values in front and expected values behind.

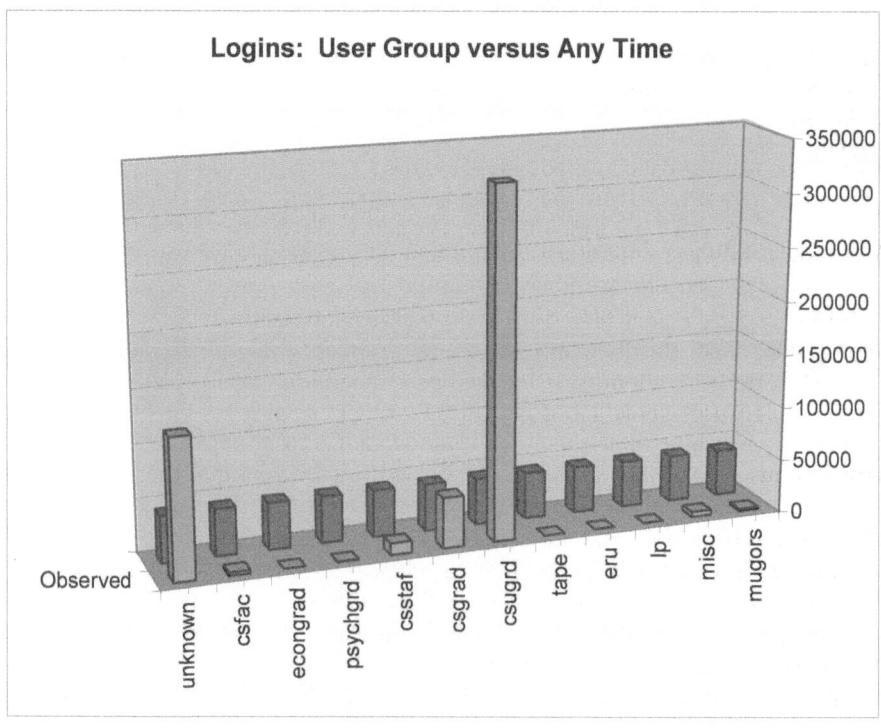

Figure 5. Graph for the (Group, Any) node for (C', U)

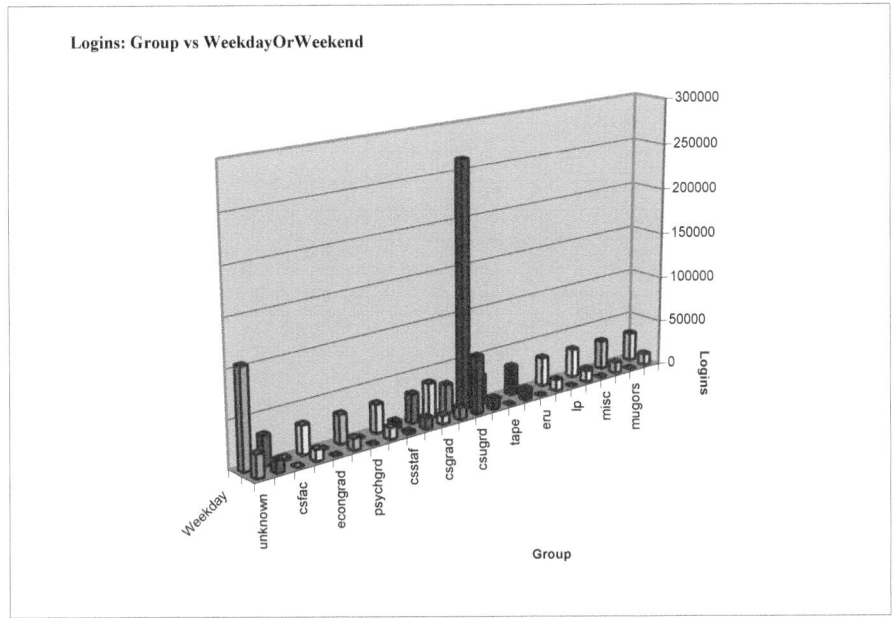

Figure 6. Graph for the (Group, WeekdayOrWeekend) node for (C', U)

The (Any, WDWE) relation is unusual because weekday logins are higher than expected based on the number of days. As well, these two factors combine to make (Group, WDWE) unusual. This relationship is shown in Figure 6 for (Group, WeekdayOrWeekend), where the X axis (across) gives the group, the Y axis (upward) gives the number of logins, and the Y axis gives the part of the week. On the X axis, the first value is the observed number of logins for the first group, the second value is the expected number of logins for this group, the third value is the observed number for the second group, the fourth value is the expected number for the second group, and so forth.

In Figure 6, the most salient feature is the great disparity between the number of logins among various groups. This disparity was addressed directly by the (C', U') distribution, where the U' distribution weighted the three groups CS undergraduates, unknown, and CS graduates much more highly than other groups.

USER	LOGINTIME	#Tuples	Coverage	Variance
Any	WeekdayOrWeekend	2	1	0.0177
Any	AcademicYearAndSem	7	1	0.0020
Any	AcademicYear	3	1	0.0015
Any	YYYY	3	1	0.0015
Any	FiscalYearAndQuarter	9	1	0.0013
Any	WeekdayName	7	1	0.0011
Any	FiscalYear	3	1	0.0011
Group	FiscalYear	29	0.806	0.0008
Any	MM	12	1	0.0007
Group	AcademicYear	29	0.806	0.0007

Table 3. Top Ranked Nodes for Hercules data and (C', U')

The ranking of the nodes for Hercules with the (C', U') distribution differs from that of (C', U). Some nodes, most notably (Group, Any) and (Group, WDWE), have lower rankings with (C', U') because of the added knowledge about the distribution in logins among the groups. As with (C, U), the top ranked node is (Any, WDWE), but this time it is because significant differences remain in the number of logins on weekdays versus weekends. As displayed in Figure 7, the graph for the (Any, AcademicYearAndSemester) node shows that fewer logins occur during the spring/summer semesters (e.g., 1999-2) than during the other semesters. In this graph, the 1998-2 and 2000-2 semesters have smaller expected values because the data cover only parts of these periods. As well, the expected values for the other semesters vary slightly because they are composed of differing numbers of days. The relationship identified as of interest, though, is that some semesters are considerably higher than expected and others lower.

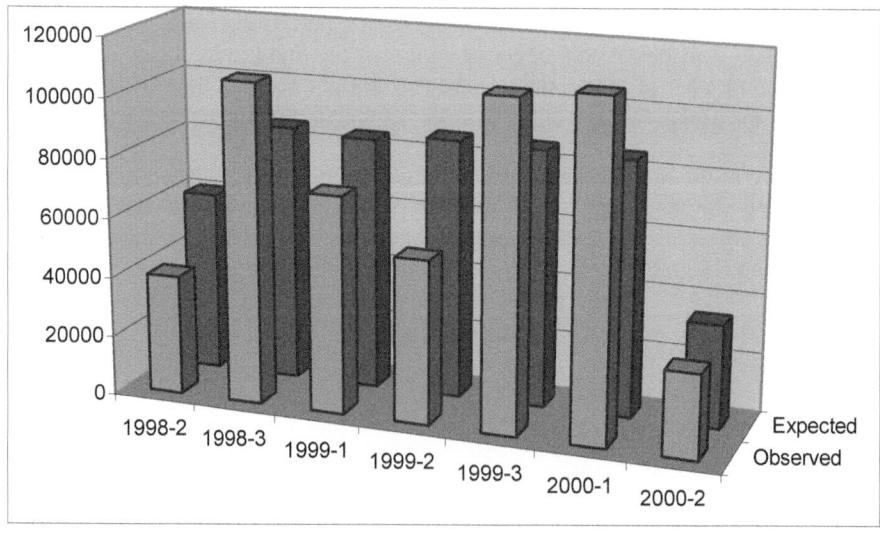

Figure 7. Graph for the (Any, AcademicYearAndSemester) Node for (C', U')

When variance is used as the measure of interest, our technique provides an easy way for a user to construct a hierarchical statistical model based on his/her knowledge of the domain. By studying the summaries correspond to the highest ranked nodes, the user may gradually recognize the factors that contribute to the observed variance. As expectations are adjusted, the variance may be reduced closer and closer to zero. Unlike traditional hierarchical statistical models, our approach allows multiple paths through the hierarchy. Although the example shown used variance, any other appropriate interest measure, such as the other 15 measures included in the HMI set of heuristic interestingness measures [10], could be used in a similar fashion.

6. Conclusions

We have specified the components of a probabilistic domain generalization graph suitable for a calendar attribute. Four types of generalization were discussed. The calendar DGG is adaptable, allowing the user to easily add new arcs and nodes to the DGG when knowledge about a calendar attribute can be expressed in different ways. The proposed DGG does not depend on special features of a particular database or database platform; thus, it can be transported to other databases. Probabilistic DGGs are appropriate for data mining. Because of the complexity of the calendar DGG, it is useful to specify distributions of values at various nodes and explore the consequences of these distributions on the interestingness ratings of various summaries of the same data.

The use of other interestingness measures besides variance should be explored in the context of providing additional guidance to the user when selecting among many possible summaries.

Acknowledgement: The research reported in this paper was supported in part by grants from the Natural Sciences and Engineering Research Council of Canada and the Institute for Robotics and Intelligent Systems. The referees contributed valuable suggestions.

References

1. A. Andrusiewicz and M. E. Orlowska, "On Granularity Factors That Affect Data Mining," *Eighth International Database Workshop, Data Mining, Data Warehousing and Client/Server Databases*, Hong Kong, 1997.
2. C. Bettini, S. Jajodia, and X.S. Wang. *Time Granularities in Databases, Data Mining, and Temporal Reasoning*. Springer-Verlag, Berlin, 2000.
3. C.L. Carter and H.J. Hamilton. Efficient attribute-oriented algorithms for knowledge discovery from large databases. *IEEE Transactions on Knowledge and Data Engineering*, 10(2):193-208, March/April 1998.
4. C. Combi, F. Pinciroli, and G. Pozzi. Managing Time Granularity of Narrative Clinical Information: The Temporal Data Model TIME-NESIS. In *Proceedings of the Third International Workshop on Temporal Representation and Reasoning (TIME-96)*, pages 88-93, Key West, Florida, May 1996.

5. D. Cukierman and J. Delgrande. A language to express time intervals and repetition. In *Proceedings of the Second International Workshop on Temporal Representation and Reasoning (TIME-95)*, pages 41-48, Melbourne, Florida, April 1995.
6. J. Euzenat. An algebraic approach to granularity in time representation. In *Proceedings of the Second International Workshop on Temporal Representation and Reasoning (TIME-95)*, pages 147-154, Melbourne, Florida, April 1995.
7. H.J. Hamilton, R.J. Hilderman, and N. Cercone. Attribute-oriented induction using domain generalization graphs. In *Proceedings of the Eighth IEEE International Conference on Tools with Artificial Intelligence (ICTAI'96)*, pages 246-253, Toulouse, France, November 1996.
8. J. Han, Y. Cai, and N. Cercone. Data-driven discovery of quantitative rules in relational databases. *IEEE Transactions on Knowledge and Data Engineering*, 5(1):29-40, February 1993.
9. R. J. Hilderman, H. J. Hamilton, and N. Cercone, Data Mining in Large Databases using Domain Generalization Graphs, *Journal of Intelligent Information Systems*, vol. 13, pp. 195-234, 1999.
10. R.J. Hilderman and H.J. Hamilton, Heuristic Measures of Interestingness. In *Proceedings of the Third European Conference on the Principles of Data mining and Knowledge Discovery (PKDD'99)*, pages 232-241, Prague, Czech Republic, September 1999.
11. J. Hobbs. Granularity. *Proc. International Joint Conference on Artificial Intelligence*, Los Angles, pages 432-435.
12. I. Merlo, E. Bertino, E. Ferrari, S. Gadia, G. Guerrini. Querying Multiple Temporal Granularity Data. In *Proceedings of the Seventh International Workshop on Temporal Representation and Reasoning (TIME-2000)*, pages 103-114, Cape Breton, Nova Scotia, Canada, July 2000.
13. C. P. Rainsford and J. F. Roddick. Adding Temporal Semantics to Association Rules, in *Third European Conference on Principles and Practice of Knowledge Discovery in Databases (PKDD'99)*. Prague: Springer, 1999, pp. 504-509.
14. D. J. Randall, H. J. Hamilton, and R. J. Hilderman. A Technique for Generalizing Temporal Durations in Relational Databases. In *Eleventh International FLAIRS Conference (FLAIRS-98)*, pages 193-197, Sanibel Island, FL, May 1998.
15. D. J. Randall, H. J. Hamilton, and R. J. Hilderman. Temporal Generalization with Domain Generalization Graphs, *International Journal of Pattern Recognition and Artificial Intelligence*. **13**(2):195-217, 1999.
16. P. Terenziani. Reasoning about Periodic Events. *In Proceedings of the Second International Workshop on Temporal Representation and Reasoning (TIME-95)*, pages 137-144, Melbourne, Florida, April 1995.
17. S.-C. Yoon and E. K. Park. An Approach to Intensional Query Answering at Multiple Abstraction Levels using Data Mining Approaches, *32nd Annual Hawaii International Conference on Systems.*

AUTOCLUST+: Automatic Clustering of Point-Data Sets in the Presence of Obstacles

Vladimir Estivill-Castro and Ickjai Lee

Department of Computer Science & Software Engineering
The University of Newcastle
Callaghan, NSW 2308, Australia
{vlad, ijlee}@cs.newcastle.edu.au

Abstract. Wide spread clustering algorithms use the Euclidean distance to measure spatial proximity. However, obstacles in other GIS data-layers prevent traversing the straight path between two points. AUTOCLUST+ clusters points in the presence of obstacles based on Voronoi modeling and Delaunay Diagrams. The algorithm is free of user-supplied arguments and incorporates global and local variations. Thus, it detects high-quality clusters (clusters of arbitrary shapes, clusters of different densities, sparse clusters adjacent to high-density clusters, multiple bridges between clusters and closely located high-density clusters) without prior knowledge. Consequently, it successfully supports correlation analyses between layers (requiring high-quality clusters) and more general locational optimization problems in the presence of obstacles. All this within $O(n \log n + [m + R] \log n)$ expected time, where n is the number of data points, m is the number of line-segments that determine the obstacles and R is the number of Delaunay edges intersecting some obstacles. A series of detailed performance evaluations illustrates the power of AUTOCLUST+ and confirms the virtues of our approach.

1 Introduction

Two-dimensional point-data sets model individual incidences of real world phenomena as a series of point locations $P = \{p_1, p_2, \ldots, p_n\}$, in some study region S in $\Re^2$. Clustering divides such a set of point-data into smaller homogeneous groups due to spatial proximity. It is central to data mining in GIS for mining massive data [2, 3, 15, 21, 33], exploring correlation between themes [5, 17, 18, 29] and exploratory data analysis [6, 7]. The information revealed from clustering may suggest hypotheses for further explorations. In particular, clustering provides indicative information (Type I information) and suggests areas of exploration (Type II information). Type I information includes the number of clusters that reside in P (note that some clustering methods are requiring this information rather than providing it), extent, scatter and size of each cluster, location in spatial dimension, relative proximity among clusters, relative density and location in temporal dimension. Type II information includes causal factors and potential explanations based on data in other layers, cluster reasoning (why are

J.F. Roddick and K. Hornsby (Eds.): TSDM 2000, LNAI 2007, pp. 133–146, 2001.

they there) and correlation analysis (which layers exhibit correlation). Obviously, this knowledge is hidden in the data prior to any clustering. Clustering is an inference to obtain this knowledge. Therefore, clustering should not impose any assumption or/and artificial constraint on the data. The analysis must produce unbiased answers. This underpins the philosophy to "let the data speak for themselves" [24].

Typically, clustering methods use the Euclidean distance to measure spatial proximity. This assumes that there exist a straight path between any two points in P. However, this assumption is unacceptable in applications where obstacles (rivers, lakes, political boundaries or mountains) prevent traversing the straight path between two points. Thus, we should consider the Euclidean distance measured along the shortest feasible path.

In this paper, we present an algorithm, AUTOCLUST+, for clustering in the presence of obstacles. The algorithm is based on AUTOCLUST [7]. The user does not have to supply arguments (like density thresholds, cut-off values or number of groups). The values of potential arguments are derived from the analysis of Delaunay Diagrams (these remove ambiguities from Delaunay Triangulations when co-circularity occurs). AUTOCLUST+ obtains all Type I information without biases or prior knowledge. It works in the presence of obstacles. It is very robust to noise, outliers and bridges. It not only detects clusters of arbitrary shapes, but also clusters of different densities. Further, it successfully finds sparse clusters near to high-density clusters, clusters linked by multiple bridges and closely located high-density clusters. These are situations in which other clustering methods typically fail [7]. Thus, AUTOCLUST+ suggests high-quality clusters to explore Type II information. All this is performed within $O(n \log n + [m + R] \log n)$ expected time, where n is the number of data points, m is the number of line-segments that determine the obstacles and R is the number of Delaunay edges intersecting some obstacle. Moreover, if the set of obstacles changes, the algorithm recomputes the clustering in only $O(m_1 \log n)$ expected time where m_1 is the number of line segments defining the new set of obstacles.

This paper is organized as follows. Section 2 introduces illustrative settings for the analysis in the presence of obstacles. Section 3 summarizes the working principle of AUTOCLUST and describes its extension for settings with obstacles. In this section we also justify the claims for the time complexity. We examine the performance of AUTOCLUST+ with a series of detailed evaluations in Section 4. These evaluations illustrate the power of our approach and confirm the virtues of the proposed method. Section 5 offers final remarks.

2   Obstacles Emerge in Many Settings

2.1   Correlation Analysis when Obstacles Exist

Typically, GIS integrates many different data-layers. Thus, finding interesting patterns in one layer is the starting point for further analysis [6]. If clustering is used for extracting point patterns in one layer towards correlation analysis between themes, it must obtain high-quality clusters. Otherwise, spatial correlation between themes is undetected. Because information is available by overlaying with other layers clustering can incorporate this information when analyzing a target layer. The direct way to incorporate other layers is to overlay them as obstacle layers that block interactions among points on the target layer. Fig. 1

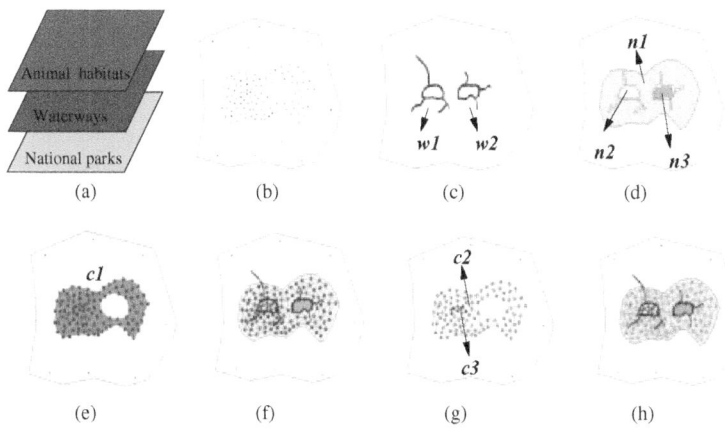

Fig. 1. Impact on clustering results when incorporating obstacles: (a) Three GIS layers; (b) Animal habitats ($n = 113$); (c) Obstacles (two waterways across national parks); (d) Three national parks; (e) Clustering result ignoring obstacles; (f) Overlay national parks and waterways layers with clustering results ignoring obstacles; (g) Clustering result accounting obstacles; (h) Overlay national parks layer with clustering result accounting obstacles.

illustrates the enhanced analysis quality of clustering and interpretability by accounting for an obstacle layer. Fig. 1(a) shows three GIS layers. Animal habitats and waterways layers are given in Fig. 1(b) and Fig. 1(c), respectively. Fig. 1(d) shows three national parks. Two of them (island-like national parks $n2$ and $n3$) are surrounded by waterways. Analyzing animal habitats using clustering without accounting for other layers as obstacles is as follows.

1. Cluster the animal habitats layer.
2. Overlay clustering result with the waterways layer to check for correlation.
3. Overlay clustering result with the national parks to check for correlation.
4. Overlay clustering result with both, waterways and national parks layers, to check correlation.

Alternatively, the other layers can be incorporated as obstacles.

1. Cluster animal habitats layer accounting for waterways layer objects as obstacles.
2. Overlay clustering result with national parks layer to find correlation.

Finding correlations suggest hypotheses for further investigations. If we take the first approach, then we will get one cluster ($c1$ as depicted in Fig. 1(e)). It is very unlikely that any interesting correlation will be found. Fig. 1(f) demonstrates the difficulty. On the other hand, we are going to have two clusters ($c2$ and $c3$ as illustrated in Fig. 1(g)) if we follow the second approach. Now, it is relatively easy to find correlations such as correlations between cluster $c2$ and national park $n1$ and between cluster $c3$ and national park $n2$. These correlations will be discovered by the system as associations and suggest possible hypotheses for further investigations. For example, a new avenue of investigation is: "why animal habitat is not found in national park $n3$ despite it is in $n2$?".

Partitioning algorithms [5, 21, 30] are not well suited for correlation analysis, since they produce convex-shaped clusters. Also, large discrepancies in size or density cause big clusters to pass undetected since their points are assigned to many meaningless fragments. Thus, these clustering methods fail to detect clusters of arbitrary shapes and clusters of different sizes. Similarly, density-based algorithms [3, 23] and grid-based algorithms [28, 31] do not serve correlation analysis well because they miss clusters of different densities and because they fail to detect sparse clusters near to high-density clusters. Graph-based algorithms [2, 6, 7, 14, 15, 32] need to overcome their vulnerability to multiple bridges.

2.2   Facility Location in the Presence of Obstacles

Partitioning algorithms aim at segmenting the n data points into k clusters. These algorithms start with a partition into k clusters and then iteratively minimize a criterion function. Typically, clusters are encoded by representatives and points belong to the nearest representatives (typically either the mean or median). This optimization corresponds to spatial facility location problems [8]. The challenge is to determine the location of points (facilities) so that the cost to the nearest points is minimized [22]. Thus, partitioning algorithms can be used for locational optimization problems. Tung *et al.* [30] suggest a modification of existing partitioning algorithm for clustering in the presence of obstacles. They use CLARANS, which embodies a randomized heuristics to solve the p-median problem, under the assumption that the number of clusters p is known (the facility location literature uses p for the number of representatives, but here $p = k$). Searching for p (or k) multiplies the complexity for the potential values of p and favors larger p since $p = n$ is an absolute minimum. Thus, such algorithm is useful only when the number of facilities (public libraries, public mail boxes or business branches) is known in advance. However, typical analysis may balance the cost of the number of facilities with the actual cost of the location solution. In other words, the number of facilities are decided after careful investigating of data. Consider the points in Fig. 2(a) representing the location of houses and a scenario where local government is planning to locate public mail boxes. An analysis will aim at maximizing convenience and accessibility while optimizing the number of mail boxes. In this region, a river runs across. Algorithms that do not take this into consideration (for example, plain AUTOCLUST) suggests two clusters (Fig. 2(b)). However, this is not a good choice as many people are forced

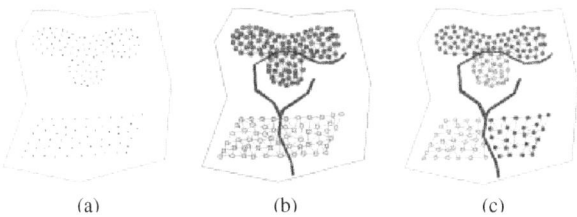

(a)   (b)   (c)

Fig. 2. Locational optimization problem in the presence of obstacles when the number of clusters is unkown: (a) Houses ($n = 155$); (b) Clustering result in the absence of obstacles (2 clusters); (c) Clustering result in the presence of obstacles (4 clusters).

to make long detours. On the other hand, AUTOCLUST+ reports 4 clusters in the region as shown in Fig. 2(c). In fact, locations of mail boxes for these four groups avoid long detours.

2.3   The Information Is in the Data

Commonly used clustering methods impose a number of assumptions and require a number of user-supplied arguments prior to clustering. The arguments required are, usually,

- the number of clusters in partitioning clustering,
- termination conditions in hierarchical clustering [13, 15, 31, 33],
- values for density thresholds in density-based clustering,
- values for resolution in grid-based clustering,
- the form of underlying probability distributions in model-based clustering [1, 9, 26] and
- threshold values for cut-off of edges in graph-based clustering.

Undoubtedly, determining suitable values for these arguments is usually a difficult task. Thus, the need to find best-fit arguments in these semi-automatic clustering demands pre-processing or/and several trial and error steps [7]. This is very expensive for massive spatial data sets. These semi-automatic algorithms inherit their argument-tuning to search best-fit arguments when they are modified for clustering in the presence of obstacles. Minimizing the need to adjust arguments not only reduces the risk of tampering with the analysis tools, but promises higher user friendliness. Our algorithm AUTOCLUST+ is able to find these values from the data automatically, so that novice users can expect high-quality clusters in the presence of obstacles. Freedom from user-supplied arguments minimizes human-generated biases and maximizes user friendliness.

3   AUTOCLUST+

Two-dimensional point-data clustering is central to spatial analysis process. It consists of non-trivial grouping of geographically close points into the same cluster. Goodchild [12] emphasized the importance of spatial analysis and modeling in which GIS is seen more as a spatial information science than a technology. The modeling and structuring of the geo-referenced data is strongly related to the analysis functionality of GIS [25]. Thus, modeling, structuring and spatial analysis are inter-related and play major roles for the success of informative mining of spatial data.

Being conscious of the interplay between modeling and analysis capability, AUTOCLUST [7] uses Voronoi Diagram and its dual Delaunay Diagram as the data model and data structure, respectively. The model, Voronoi Diagram, overcomes the limitations of conventional vector and raster models and guarantees unique modeling of discrete point-data [6, 7, 10, 11]. The structure, Delaunay Diagram, is succinct and efficient for clustering [2, 5, 6, 7, 14].

Clustering finds global sudden changes in point density. In Delaunay Diagrams points in the border of clusters tend to have greater standard deviation of length of their incident edges. This is because border points have both short edges and long edges. The short edges connect points within a cluster while the long edges straddle between clusters or between a cluster and noise points. AUTOCLUST takes into account local and global variations to ensure that relatively homogeneous clusters are not broken into tiny uninteresting sub-sets and relatively heterogeneous sub-sets are split into meaningful clusters. This analysis allows AUTOCLUST to be argument free. It reports high-quality localized excesses.

AUTOCLUST proceeds in three phases. Each phase is an edge correction phase. The idea is to start with the Delaunay Diagram and produce a planar graph that synthesizes the discrete relation p_i is connected to p_j if and only if p_i and p_j are in the same cluster. The three-phase constitutes a process to automatically find cluster boundaries in the global aspect of view. Phase I eliminates edges that are evidently too long. In order to detect this, AUTOCLUST computes a local indicator $Local\_Mean(p_i)$ for each p_i. It also uses a global indicator $Mean\_St\_Dev(P)$. The value of $Local\_Mean(p_i)$ is the mean length of edges incident to point p_i. The global indicator $Mean\_St\_Dev(P)$ is the average of the standard deviations in the length of incident edges for all points p_i. The first phase classifies incident edges for each p_i into $Short\_Edges(p_i)$ (lengths of edges are less than $Local\_Mean(p_i)$ - $Mean\_St\_Dev(P)$), $Long\_Edges(p_i)$ (lengths of edges are greater than $Local\_Mean(p_i)$ + $Mean\_St\_Dev(P)$) and $Other\_Edges(p_i)$ (incident edges - ($Short\_Edges(p_i)$ ∪ $Long\_Edges(p_i)$)). AUTOCLUST obtains rough boundaries of clusters by removing $Long\_Edges(p_i)$ and $Short\_Edges(p_i)$ for all points p_i. $Long\_Edges(p_i)$ are the ones that are too long joining points in different clusters. Thus, they are permanently removed. On the other hand, $Short\_Edges(p_i)$ may correspond to links between points within a cluster but also may correspond to bridges between clusters. Thus, removing incident edges $e \in Short\_Edges(p_i)$ for all p_i eliminates certain types of bridges.

Phase II recuperates edges in *Short_Edges*(p_i) that are intra-cluster links. It computes connected components and re-connects isolated *singleton connected components* (connected components having only one element) to *non-trivial connected components* (connected components having more than one element) if and only if they are very close (singleton connected components p_i with $e = (p_i, p_j) \in$ *Short_Edges*(p_i) and $p_j \in$ some non-trivial connected components).

Phase III extends the notion of neighborhood of a point p_i to those points reachable by paths of length 2 or less. Then, the indicators of local variation are recalculated. The third phase detects and removes inconsistent edges for each p_i by the use of the new indicator *Local_Mean*$_{2,G}(p_i)$ (the average length of edges in paths of 2 or less edges starting at p_i). Fig. 3 illustrates the three-phase cluster detection process.

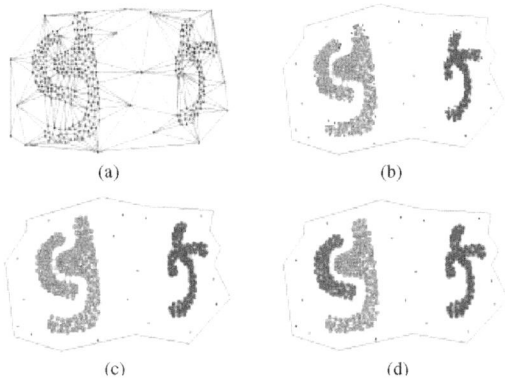

(a)   (b)

(c)   (d)

Fig. 3. AUTOCLUST operates in three-phase: (a) Delaunay Diagram ($n = 391$); (b) After Phase I (initial 2 clusters); (c) After Phase II (2 clusters with refinement); (d) After Phase III (3 clusters).

In a proximity graph, points are linked by edges if and only if they seem to be close by some proximity measure [19]. In Delaunay Diagrams, each Delaunay edge represents a relationship that indicates two end-points are relative neighbors (there is a circle empty of other points). Strong interactions (short edges) indicate two end-points belong to the same cluster while weak ones (long edges) imply end-points belong to distinct clusters. However, in the presence of obstacles, interactions (Delaunay edges) and paths are blocked by obstacles. As a consequence, Delaunay edges intersecting defining lines of obstacles no longer represent links between members of the same cluster. Thus, AUTOCLUST+ tackles obstacles from a graph-based clustering approach. AUTOCLUST+ removes all edges in the Delaunay Diagram intersecting some obstacles. Therefore, the two end-points of a Delaunay edge traversing some obstacles may or may not belong to the same cluster. The decision now purely depends on the length of edges on the detour path [1]. Fig. 4 illustrates a detour path. Thick solid lines represent the detour distance in Fig. 4(a). A detour path is shown in Fig. 4(b). Dotted edges represent Delaunay edges removed because they traverse an obstacle. Note that, the lines determining the detour distance are not relations in the Delaunay Diagram. The detour distance is approximated by existing Delaunay edges. The length of a detour path in the Delaunay Diagram is a good approximation to the detour distance because the Delaunay Diagram is a spanner [16].

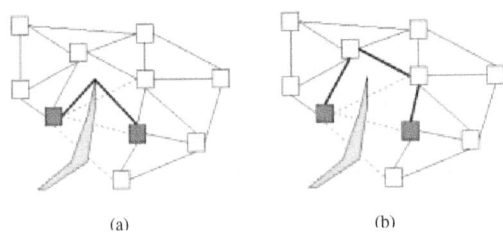

(a)   (b)

Fig. 4. Detour distance and detour path: (a) detour distance (thick solid lines); (b) detour path (thick solid lines).

The steps of AUTOCLUST+ are as follows:

1. Construct Delaunay Diagram.
2. Calculate the global variation indicator ($Mean\_St\_Dev(P)$).
3. For all edges e, remove e if it intersects some obstacles.
4. Apply the three phases of AUTOCLUST to the planar graph resulting from the previous steps.

AUTOCLUST+ constructs Delaunay Diagram as a basic data structure. We use the global indicator to obtain global information before intersecting with obstacles, since obstacles are from another data layer. The next step removes Delaunay edges traversing obstacles. Now, incident edges have new local indicators $Local\_Mean(p_i)$ and $Local\_Mean_{2,G}(p_i)$, since some of their incident edges were removed. Using the three-phase edge elimination procedure on the newly generated proximity graph provides very satisfactory results that will be illustrated in following section.

AUTOCLUST+ requires $O(n \log n)$ time to construct Delaunay Diagram. The computing of $Mean\_St\_Dev(P)$ is bounded by twice the number of edges. Thus, this takes linear time, i.e. $O(n)$ time. Removal of edges that traverse obstacles requires $O([m + R] \log n)$, where m is the number of line segments that determine the obstacles and R is the total number of edges removed from the

---

[1] The *detour path* is defined as the shortest path between two points in the Delaunay Diagram where all the edges on the path are Delaunay edges that do not traverse any obstacles.

Delaunay Diagram. This is because for each line-segment l defining an obstacle, its endpoints can be located within the Delaunay Diagram in $O(\log n)$ expected time. Then the planar structure of the Delaunay Diagram can be navigated along the line-segment l collecting edges to remove in $O(\log n)$ expected time. Our implementation uses LEDA [20]. Typically, $m \ll n$ resulting in $o(n)$ time unless the set of obstacles is extremely complex. Finally, the three-phase cleaning process works in linear time [7]. Thus, AUTOCLUST+ only requires $O(n \log n + [m + R] \log n)$ time. The time complexity of AUTOCLUST+ is dominated by computing Delaunay Diagram. This means that once we have the Delaunay Diagram, it can be stored and new clusterings can be obtained in $O(n)$ time for new sets of obstacles.

4   Performance Evaluation

This section presents the results of experiments with four synthetic data sets. We present the clusters produced by AUTOCLUST+ in the absence and in the presence of obstacles. AUTOCLUST+ in the absence of obstacles reduces to AUTOCLUST, so this illustration also reveals how the approach is powerful to identify high-quality clusters without prior knowledge. The results are of the same high-quality in the presence of obstacles. Fig. 5(a) consists of clusters of various shapes and different densities. AUTOCLUST+ correctly identifies 12 clusters in the absence of obstacles (Fig. 5(b)). Obstacles analogous to two rivers and a circle-like lake split some clusters. Thus, AUTOCLUST+ properly reports 18 clusters (Fig. 5(c)).

Fig. 5(d) contains examples where clustering algorithms typically fail. Without obstacles (Fig. 5(e)), AUTOCLUST+ correctly identifies clusters connected by multiple bridges (top-left), sparse clusters near to high-density clusters (top-right), closely located high-density clusters (bottom-left) and doughnut-like and island-like clusters (bottom-right). When obstacles are considered (Fig. 5(f)), AUTOCLUST+ correctly identifies 14 clusters. AUTOCLUST+ has no problem with CHAMELEON's benchmark [15] data as shown in Fig. 5(g), (h) and (i). Finally, Fig. 5(j) is a CSR (complete spatial randomness) data set. Thus, there is no localized excess. If boundaries are imposed on the data set, these obstacles define five clusters. AUTOCLUST+ correctly reports these clusters, while no localized excesses are detected in the absence of obstacles.

AUTOCLUST+ has several direct improvements over COE [30]. First, AUTOCLUST+ uses the Euclidean distance and not the square of the Euclidean distance. Using the square of the Euclidean distance is typically motivated by considerations like speeding the implementation by a constant factor and/or by an inductive principle of least-squares (as is the case with k-means [4]). However, using squares of the Euclidean distance in the modeling makes the approach extremely sensitive to outliers and/or noise [5, 27]. Second, AUTOCLUST+ does not need to experiment with different values for the number of clusters. This is a significant benefit, both in CPU-time and in user-analysis time. Third, AUTOCLUST+ operates with all the data and does not require the preprocessing of

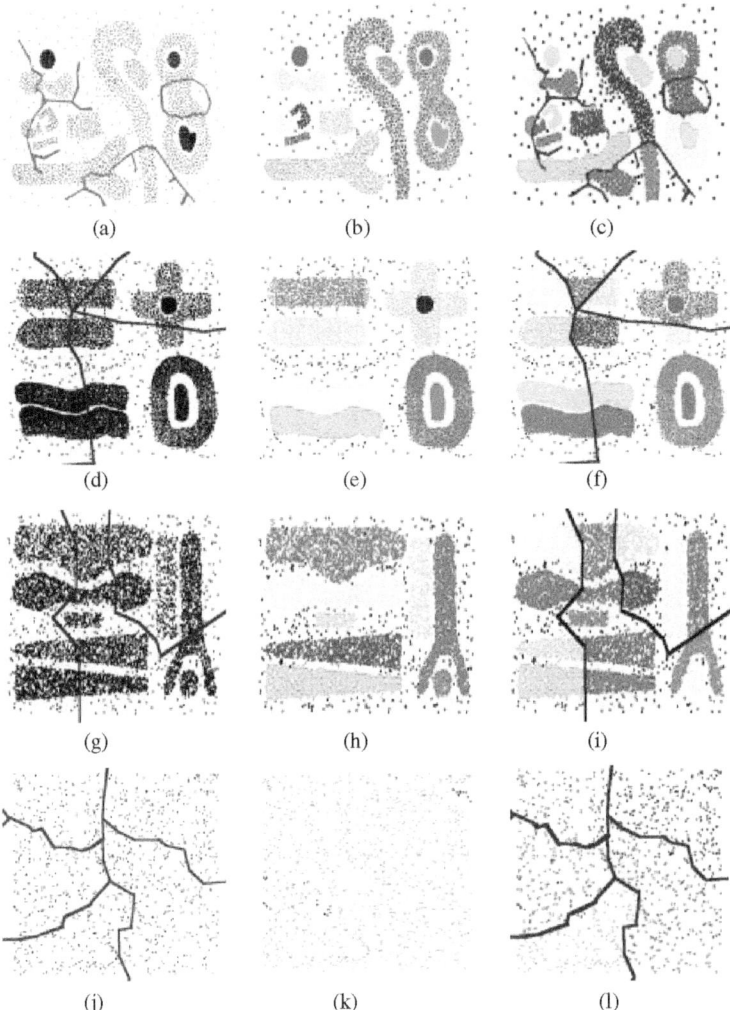

Fig. 5. Synthetic data sets: (a) Dataset I ($n = 3250$ with 2 rivers and a lake); (b) Clustering result of Dataset I in the absence of obstacles (12 clusters); (c) Clustering result of Dataset I in the presence of obstacles (18 clusters); (d) Dataset II ($n = 8000$ with a range of mountains); (e) Clustering result of Dataset II in the absence of obstacles (8 clusters); (f) Clustering result of Dataset II in the presence of obstacles (14 clusters); (g) Dataset III ($n = 8000$ with 2 rivers); (h) Clustering result of Dataset III in the absence of obstacles (8 clusters); (i) Clustering result of Dataset III in the presence of obstacles (15 clusters); (j) Dataset IV ($n = 1000$ with political boundaries); (k) Clustering result of Dataset IV in the absence of obstacles (1 cluster); (l) Clustering result of Dataset IV in the presence of obstacles (5 clusters).

COE to form mini-clusters. This construction of mini-clusters introduces numerical error and biases the grouping. Thus, the quality of the clustering is reduced and the exploratory capability is limited in COE. Fourth, COE uses a random-

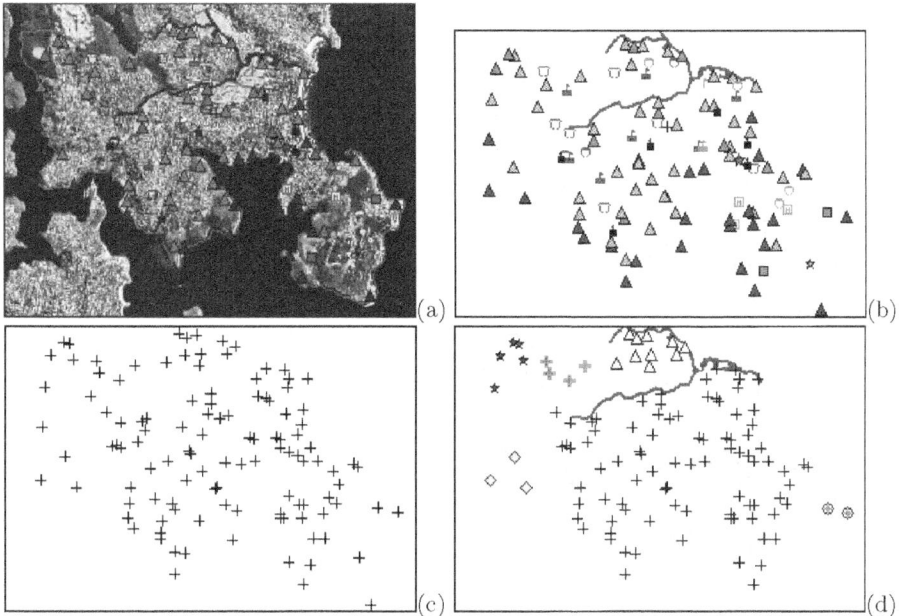

Fig. 6. Illustration with real data set from Sydney, Australia: (a) The region of study has rivers and sorounded by the ocean. (b) Clustering is to be performed for point features in the presence of obstacles. (c) If obstacles are ignored, one cluster is obtained involving all points. (d) If obstacles are considered, 6 clusters are found and 4 objects are isolated.

ized interchange heuristic to approximate the optimum set of k representatives for the clusters. This heuristics is a very poor hill-climber that can not guarantee even local optimality and that sacrifices significantly the quality of the clustering for speed in the optimization. Much better variants of interchange heuristics are well know from the literature of facility location, and they have been applied to medoid-based clustering [8]. Fifth, because of the representative-based nature of COE, the algorithm has a bias for convex clusters, and in particular for spherical clusters. This is probably acceptable when statistical justifications allow to assume that the data has a probabilistic distribution from a mixture model. However, in the presence of obstacles, this assumption will certainly be very difficult to sustain. AUTOCLUST+ is not restricted to convex clusters. In fact, Fig. 6 illustrates some of AUTOCLUST+ advantages with a real data set. In particular, AUTOCLUST+ detects that several edges of the Delauney are to long, and even without obstacles, although it produces only one cluster, its shape is not convex and the bays around the data set are identified.

One thing in which AUTOCLUST+ and COE are similar is that they are a "tightly-coupled" [30] approach to including obstacles into the clustering. That is, rather than computing the clustering ignoring the obstacles and then cutting the resulting groups by the obstacles, the obstacles are used to modify parameters (the distance) when considering the cluster. In the case of COE, there is no real explanation to avoiding the "loosely-coupled" approach, although it seems rather more efficient but producing the same results as the "tightly-coupled" approach. On the other hand, in the case of AUTOCLUST+, the computational efficiency is essentially the same, and both approaches could be taken. However, in AUTOCLUST+ we recommend the "tightly-coupled" approach in that the evaluation of global effects is considered with all Delauney edges, even those that intersect obstacles, since these is what allows the discovery of density estimates over the region of study. But when analyzing the neighbors of a data point, obstacles are considered, since and edge intersecting an obstacle is not an edge that makes its corresponding end-points neighbors, but precisely the opposite.

As was pointed out before, Fig. 6 illustrates AUTOCLUST+ on a real-data sets. Note that the data-points above the river that cuts East-West are actually split into two cluster because the river has an arm that almost circles, although it does not disconnect such data points.

5   Final Remarks

Accurate data analysis produces better decision making. GIS integrates several data layers and their inter-relation influences the clustering of one. In this paper, we incorporate the information of other layers, particularly obstacles, to a target layer to produce accurate clustering results.

AUTOCLUST+ considers both global variation and local variation indicators in order not to split homogeneous clusters into small meaningless sub-sets and in order to divide heterogeneous sub-sets into meaningful clusters. AUTOCLUST+ is free of user-supplied arguments and work uniformly well in the absence and in the presence of obstacles. Thus, it helps users to scrutinize data sets for locational optimization problem with obstacles. It allows fast exploration with respect to several sets of obstacles. Further, novice users can expect high-quality of clustering results when analyst experts are not available, which eventually maximize user friendliness and minimize human-generated bias. AUTOCLUST+ uses only linear time to produce high-quality clusters (clusters of arbitrary shapes, clusters of different densities, sparse clusters adjacent to high-density clusters, multiple bridges between clusters and closely located high-density clusters) in the presence of various sets of obstacles when the Delaunay Diagram is available. Therefore, it serves correlation analysis well with fast exploration time.

References

[1] J. D. Banfield and A. E. Raftery. Model-based Gaussian and Non-Gaussian Clustering. *Biometrics*, 49:803–821, 1993.

[2] C. Eldershaw and M. Hegland. Cluster Analysis using Triangulation. In B. J. Noye, M. D. Teubner, and A. W. Gill, editors, *Computational Techniques and Applications: CTAC97*, pages 201–208. World Scientific, Singapore, 1997.

[3] M. Ester, M. P. Kriegel, J. Sander, and X. Xu. A Density-Based Algorithm for Discovering Clusters in Large Spatial Databases with Noise. In *Proceedings of the 2nd International Conference on Knowledge Discovery and Data Mining*, pages 226–231, 1996.

[4] V. Estivill-Castro. Hybrid genetic algorithms are better for spatial clustering. In *Proc. 6th Pacific Rim Intern. Conf. Artificial Intelligence PRICAI 2000*, Melbourne, 2000. Springer-Verlag Lecture Notes in AI, to appear.

[5] V. Estivill-Castro and M. E. Houle. Robust Clustering of Large Geo-referenced Data Sets. In *Proceedings of the 3rd Pacific-Asia Conference on Knowledge Discovery and Data Mining*, pages 327–337, 1999.

[6] V. Estivill-Castro and I. Lee. AMOEBA: Hierarchical Clustering Based on Spatial Proximity Using Delaunay Diagram. In *Proceedings of the 9th International Symposium on Spatial Data Handling*, 2000. to appear.

[7] V. Estivill-Castro and I. Lee. AUTOCLUST: Automatic Clustering via Boundary Extraction for Massive Point-data Sets. In *Proceedings of the 5th International Conference on Geocomputation*, 2000. to appear. Extended version is available at http://www.cs.newcastle.edu.au/Dept/techrep.html as a technical report.

[8] V. Estivill-Castro and A.T. Murray. Discovering associations in spatial data - An efficient medoid based approach. In X. Wu, R. Kotagiri, and K.K. Korb, editors, *Proceedings of the 2nd Pacific-Asia Conference on Knowledge Discovery and Data Mining (PAKDD-98)*, pages 110–121, Melbourne, Australia, 1998. Springer-Verlag Lecture Notes in Artificial Intelligence 1394.

[9] C. Fraley and A. E. Raftery. How Many Clusters? Which Clustering Method? Answers Via Model-Based Cluster Analysis. *The Computer Journal*, 41(8):578–588, 1998.

[10] C. M. Gold. Problems with handling spatial data - The Voronoi approach. *CISM Journal ACSGC*, 45(1):65–80, 1991.

[11] C. M. Gold. The meaning of Neighbour. In G. Tinhofer and G. Schmidt, editors, *Theories and Methods of Spatio-Temporal Reasoning in Geographic Space*, pages 220–235, Berlin, 1992. Springer-Verlag Lecture Notes in Computer Science 639.

[12] M. F. Goodchild. Geographical information science. *International Journal of Geographical Information Systems*, 6(1):31–45, 1992.

[13] S. Guha, R. Rastogi, and K. Shim. CURE: An Efficient Clustering Algorithm for Large Databases. In *Proceedings of the ACM SIGMOD International Conference on Management of Data*, pages 73–84, 1998.

[14] I. Kang, T. Kim, and K. Li. A Spatial Data Mining Method by Delaunay Triangulation. In *Proceedings of the 5th International Workshop on Advances in Geographic Information Systems (GIS-97)*, pages 35–39, 1997.

[15] G. Karypis, E. Han, and V. Kumar. CHAMELEON: A Hierarchical Clustering Algorithm Using Dynamic Modeling. *IEEE Computer: Special Issue on Data Analysis and Mining*, 32(8):68–75, 1999.

[16] M. Keil and G Gutwin. The Delauney triangulation closely approximates the complete graph. In F. Denhe, J.-R. Sack, and N. Snatoro, editors, *Proceedings of*

the *First Workshop on Algorithms and Data Structures WADS-89*, pages 47–56. Springer-Verlag Lecture Notes in Computer Science 382, 1989.

[17] E. M. Knorr, R. T. Ng, and D. L. Shilvock. Finding Aggregate Proximity Relationships and Commonalities in Spatial Data Minings. *IEEE Transactions on Knowledge and Data Engineering*, 8(6):884–897, 1996.

[18] E. M. Knorr, R. T. Ng, and D. L. Shilvock. Finding Boundary Shape Matching Relationships in Spatial Data. In *Proceedings of the 5th International Symposium on Spatial Databases*, pages 29–46, Berlin, 1997. Springer-Verlag Lecture Notes in Computer Science 1262.

[19] G. Liotta. Low Degree Algorithm for Computing and Checking Gabriel Graphs. Technical Report 96-28, Department of Computer Science, Brown University, 1996.

[20] K. Mehlhorn and S. Näher. *LEDA A platform for combinatorial and geometric computing*. Cambridge University Press, Cambridge, 1999.

[21] R. T. Ng and J. Han. Efficient and Effective Clustering Method for Spatial Data Mining. In *Proceedings of the 20th International Conference on Very Large Data Bases (VLDB)*, pages 144–155, 1994.

[22] A. Okabe, B. N. Boots, and K. Sugihara. *Spatial Tessellations: Concepts and Applications of Voronoi Diagrams*. John Wiley & Sons, West Sussex, 1992.

[23] S. Openshaw. A Mark 1 Geographical Analysis Machine for the automated analysis of point data sets. *International Journal of GIS*, 1(4):335–358, 1987.

[24] S. Openshaw. Two exploratory space-time-attribute pattern analysers relevant to GIS. In S. Fotheringham and P. Rogerson, editors, *Spatial Analysis and GIS*, pages 83–104. Taylor and Francis, London, 1994.

[25] J. F. Paper and D. J. Maguire. Design models and functionality in gis. *Computers and Geosciences*, 18(4):387–394, 1992.

[26] C. Posse. Hierarchical Model-Based Clustering for Large Datasets. Technical Report 363, Department of Statistics, University of Washington, 1999.

[27] P. J. Rousseeuw and A. M. Leroy. *Robust Regression and Outlier Detection*. John Wiley, NY, 1987.

[28] E. Schikuta and M. Erhart. The BANG-Clustering System: Grid-Based Data Analysis. In X. Liu, P. Cohen, and M. Berthold, editors, *Proceedings of the Second International Symposium IDA-97*, pages 513–524, Berlin, 1997. Advances in Intelligent Data Analysis, Springer-Verlag Lecture Notes in Computer Science 1280.

[29] E. Son, I. Kang, and K. Li. A Spatial Data Mining Method by Clustering Analysis. In *ACM-GIS 1998*, pages 157–158, 1998.

[30] A. K. H. Tung, J. Hou, and J. Han. COE: Clustering with Obstacles Entities, A Preliminary Study. In *Proceedings of the 4th Pacific-Asia Conference on Knowledge Discovery and Data Mining*, 2000.

[31] W. Wang, J. Yang, and R. Muntz. STING+: An Approach to Active Spatial Data Mining. In *Proceedings of the International Conference on Data Engineering*, pages 116–125, 1999.

[32] C. T. Zahn. Graph-Theoretical Methods for Detecting and Describing Gestalt Clusters. *IEEE Transactions of Computers*, 20(1):68–86, 1971.

[33] T. Zhang, R. Ramakrishnan, and M. Livny. BIRCH: An Efficient Data Clustering Method for Very Large Databases. In *Proceedings of the ACM SIGMOD International Conference on Management of Data*, pages 103–114, 1996.

An Updated Bibliography of Temporal, Spatial, and Spatio-temporal Data Mining Research

John F. Roddick[1], Kathleen Hornsby[2], and Myra Spiliopoulou[3]

[1] School of Informatics and Engineering,
Flinders University of South Australia,
PO Box 2100, Adelaide 5001, South Australia.
roddick@cs.flinders.edu.au
[2] National Centre for Geographic Information and Analysis,
University of Maine,
Orono, Maine, ME 04469-5711, USA.
khornsby@spatial.maine.edu
[3] Institut für Technische und Betriebliche Informationssysteme
Otto-von-Guericke Universität Magdeburg
Universitätsplatz 2, D-39016 Magdeburg, Germany.
myra@iti.cs.uni-magdeburg.de

Abstract. Data mining and knowledge discovery have become important issues for research over the past decade. This has been caused not only by the growth in the size of datasets but also in the availability of otherwise unavailable datasets over the Internet and the increased value that organisations now place on the knowledge that can be gained from data analysis. It is therefore not surprising that the increased interest in temporal and spatial data has led also to an increased interest in mining such data. This bibliography subsumes an earlier bibliography and shows that the value of investigating temporal, spatial and spatio-temporal data has been growing in both interest and applicability.

Keywords: Temporal Data Mining, Temporal Rules, Temporal Patterns, Spatial Data Mining, Spatio-Temporal Data Mining.

1 Introduction

Temporal and spatial data mining have grown rapidly as an exciting subfield of data mining. There are many reasons for this –

- A growth in the volume of data being collected and requiring analysis,
- An increase in the general availability of data through the Internet and as a result of electronic commerce and inter-enterprise applications,
- A recognition in the value and commercial advantage that the results of data mining can bring,
- The recognition that the temporal and spatial context of data is special and needs to be explicitly accommodated.

J.F. Roddick and K. Hornsby (Eds.): TSDM 2000, LNAI 2007, pp. 147–163, 2001.

Over the past four years there has been a substantial increase (approximately 90% pa compounded), in temporal, spatial and spatio-temporal data mining publications. In particular, there has been a large volume of research reported since the previous bibliography was published in *SIGKDD Explorations* (listed here in Section 2.9). This bibliography subsumes and reorganises this earlier version.

This bibliography has been subdivided into two major sections. The first contains those publications directly relevant to the field, partitioned into various categories. The second section lists conferences from which the publications originated and thus could be consulted for future papers. This latter section is useful rather than exhaustive.

The publications in Section 2 are listed in alphabetical order by first author and placed into categories as follows:

- **Frameworks**. This category lists those papers dealing primarily with models for spatial and temporal knowledge discovery. However, we also include here papers that discuss new ways of viewing data mining activity relevant to this bibliography.
- **Temporal and Spatial Association Rule Mining**. This category combines all papers that contribute to the problem of discovering association rules from temporal or spatial data.
- **Discovery of Temporal Patterns**. This research is concerned with the discovery of patterns or trends over time. The data need not itself be temporal but an ordering is required.
- **Time Series Mining**. This category includes research into the occurrence of events over time.
- **Discovery of Causal and/or Temporal Rules**. This category lists those papers that search for temporal relationships between (sets of) events. Also included here are also a number of papers that do not easily categorise themselves under other headings but deserve to be included in a bibliography such as this.
- **Spatial Data Mining**. This section lists those papers pertain to spatial and geo-referenced data mining.
- **Spatial and Spatio-Temporal Clustering Techniques**. This category includes those publications that propose algorithms or frameworks for spatial and spatio-temporal clustering.
- **Spatio-Temporal Data Mining**. This category contains those papers that explicitly accommodate the special semantics of both space and time.
- **Theses, Surveys, Books and Previous Bibliographies**. While these often deal with more than one issue, they also tend to have a broader scope and be more comprehensive in their treatment of an issue. We have therefore listed these works separately.

Some papers provide useful contributions in more than one area and have therefore been listed under more than one category. For completeness, the publications in this volume are listed.

Note that as the bibliography aims to include only papers that directly advance the process of finding temporal and/or spatial rules about data, it does not, for example, include papers describing the incremental maintenance of rulesets of association rules (albeit that this may take place over time).

This final section is a list of conferences that, by virtue of their past selection of papers, are likely to provide relevant research in the future.

2   Publications

2.1   Frameworks for Temporal, Spatial, and Spatio-temporal Mining

Al-Naemi, S. (1994). *A Theoretical Framework for Temporal Knowledge Discovery*. In Proc. International Workshop on Spatio-Temporal Databases, Benicassim, Spain, 23-33.

Berger, G. and Tuzhilin, A. (1998). Discovering unexpected patterns in temporal data using temporal logic. In *Temporal Databases - Research and Practice*. O. Etzion, S. Jajodia and S. Sripada, Eds. Berlin. Springer-Verlag. Lecture Notes in Computer Science. **1399**: 281-309.

Black, M.M. and Hickey, R.J. (1999). Maintaining the Performance of a Learned Classifier under Concept Drift. *Intelligent Data Analysis* **3**(6): 453-474.

Chen, X. and Petrounias, I. (1998). *A framework for temporal data mining*. In Proc. Ninth International Conference on Database and Expert Systems Applications, DEXA'98, Vienna, Austria, G. Quirchmayr, E. Schweighofer and T. J. M. Bench-Capon, Eds., Springer-Verlag. Lecture Notes in Computer Science. **1460**: 796-805.

Fawcett, T. and Provost, F. (1999). *Activity Monitoring: Noticing Interesting Changes in Behaviour*. In Proc. Fifth International Conference on Knowledge Discovery and Data Mining, San Diego, CA, USA, S. Chaudhuri and D. Madigan, Eds., ACM Press. 53-62.

Peuquet, D.J. (1994). It's about Time: A Conceptual Framework for the Representation of Spatiotemporal Dynamics in Geographic Information Systems. *Annals of the Association of American Geographers* **84**: 441-461.

Rainsford, C.P. and Roddick, J.F. (1996). *Temporal data mining in information systems: a model*. In Proc. Seventh Australasian Conference on Information Systems, Hobart, Tasmania, **2**: 545-553.

Saraee, M.H. and Theodoulidis, B. (1995). *Knowledge discovery in temporal databases: The initial step*. In Proc. DOOD'95 Post-Conference Workshop "Knowledge Discovery in Databases and DOOD", Singapore, K. Ong, S. Conrad and T. W. Ling, Eds., 17-22.

Spiliopoulou, M. and Roddick, J.F. (2000). Higher Order Mining: Modelling and Mining the Results of Knowledge Discovery. In *Data Mining II - Proc. Second International Conference on Data Mining Methods and Databases*. N. Ebecken and C. A. Brebbia, Eds. Cambridge, UK. WIT Press. 309-320.

2.2   Temporal and Spatial Association Rule Mining

Chen, X. and Petrounias, I. (1999). *Mining Temporal Features in Association Rules*. In Proc. 3rd European Conference on Principles and Practice of Knowledge Discovery in Databases (PKDD'99), Prague, J. M. Zytkow and J. Rauch, Eds., Springer. Lecture Notes in Artificial Intelligence. **1704**: 295-300.

Chen, X., Petrounias, I. and Heathfield, H. (1998). *Discovering temporal association rules in temporal databases*. In Proc. International Workshop on Issues and Applications of Database Technology (IADT'98), 312-319.

Estivill-Castro, V. and Murray, A.T. (1998). *Discovering associations in spatial data-an efficient mediod based approach*. In Proc. Second Pacific-Asia Conference on Research and Development in Knowledge Discovery and Data Mining, PAKDD-98, Springer-Verlag, Berlin. 110-121.

Koperski, K. and Han, J. (1995). *Discovery of Spatial Association Rules in Geographic Information Databases*. In Proc. Fourth International Symposium on Large Spatial Databases, Maine, 47-66.

Rainsford, C.P. and Roddick, J.F. (1999). *Adding Temporal Semantics to Association Rules*. In Proc. 3rd European Conference on Principles and Practice of Knowledge Discovery in Databases (PKDD'99), Prague, J. M. Zytkow and J. Rauch, Eds., Springer. Lecture Notes in Artificial Intelligence. **1704**: 504-509.

Ye, S. and Keane, J.A. (1998). *Mining association rules in temporal databases*. In Proc. International Conference on Systems, Man and Cybernetics, IEEE, New York. 2803-2808.

2.3   Discovery of Temporal Patterns

Agrawal, R., Lin, K.-I., Sawhney, H.S. and Shim, K. (1995). *Fast similarity search in the presence of noise, scaling, and translation in time-series databases*. In Proc. Twenty-First International Conference on Very Large Data Bases, Zurich, Switzerland, Morgan Kaufmann. 490-501.

Agrawal, R. and Srikant, R. (1995). *Mining sequential patterns*. In Proc. Eleventh International Conference on Data Engineering, Taipei, Taiwan, P. S. Yu and A. S. P. Chen, Eds., IEEE Computer Society Press. 3-14.

Bayardo Jr, R.J. (1998). *Efficiently mining long patterns from databases*. In Proc. ACM SIGMOD Conference on the Management of Data, Seattle, WA, USA, ACM Press. 85-93.

Berger, G. and Tuzhilin, A. (1998). Discovering unexpected patterns in temporal data using temporal logic. In *Temporal Databases - Research and Practice*. O. Etzion, S. Jajodia and S. Sripada, Eds. Berlin. Springer-Verlag. Lecture Notes in Computer Science. **1399**: 281-309.

Bettini, C., Wang, X.S., Jajodia, S. and Lin, J.-L. (1998). Discovering Frequent Event Patterns with Multiple Granularities in Time Sequences. *IEEE Transactions on Knowledge and Data Engineering* **10**(2): 222-237.

Chen, M.S., Park, J.S. and Yu, P.S. (1996). *Data mining for path traversal patterns in a web environment*. In Proc. Sixteenth International Conference on Distributed Computing Systems, 385-392.

Dietterich, T.G. and Michalski, R.S. (1985). Discovering patterns in sequences of events. *Artificial Intelligence* **25**: 187-232.

Han, J., Gong, W. and Yin, Y. (1998). *Mining segment-wise periodic patterns in time-related databases*. In Proc. Fourth International Conference on Knowledge Discovery and Data Mining, AAAI Press, Menlo Park. 214-218.

Klösgen, W. (1995). *Deviation and association patterns for subgroup mining in temporal, spatial, and textual data bases*. In Proc. First International Conference on Rough Sets and Current Trends in Computing, RSCTC'98, Springer-Verlag, Berlin,. 1-18.

Lesh, N., Zaki, M.J. and Ogihara, M. (1999). *Mining Features for Sequence Classification*. In Proc. Fifth ACM SIGKDD International Conference on Knowledge Discovery and Data Mining, San Diego, S. Chaudhuri and D. Madigan, Eds., ACM Press.

Li, Y., Wang, X.S. and Jajodia, S. (2000). *Discovering Temporal Patterns in Multiple Granularities*. In Proc. International Workshop on Temporal, Spatial and Spatio-Temporal Data Mining, TSDM2000, Lyon, France, J. F. Roddick and K. Hornsby, Eds., Springer. Lecture Notes in Artificial Intelligence. **2007**.

Lin, D.-I. and Kedem, Z.M. (1998). *Pincer Search: A new algorithm for discovering the maximum frequent set*. In Proc. International Conference on Extending Database Technology, EDBT'98, Valencia, Spain, 385-392.

Mannila, H. and Toivonen, H. (1996). *Discovering generalised episodes using minimal occurences*. In Proc. Second International Conference on Knowledge Discovery and Data Mining (KDD-96), Portland, Oregon, AAAI Press, Menlo Park. 146-151.

Mannila, H., Toivonen, H. and Verkamo, A.I. (1995). *Discovering frequent episodes in sequences*. In Proc. First International Conference on Knowledge Discovery and Data Mining (KDD-95), Montreal, Quebec, Canada, U. Fayyad, M. and R. Uthurusamy, Eds., AAAI Press, Menlo Park, CA, USA. 210-215.

Ong, K.L., Conrad, S. and Ling, T.W., Eds. (1995). *Knowledge Discovery and Temporal Reasoning in Deductive and Object-Oriented Databases*. Proceedings of the DOOD'95 Post-Conference Workshops on Integration of Knowledge Discovery in Databases with Deductive and Object-Oriented Databases (KDOOD) and Temporal Reasoning in Deductive and Object-Oriented Databases (TDOOD). Singapore.

Padmanabhan, B. and Tuzhilin, A. (1996). *Pattern discovery in temporal databases: a temporal logic approach*. In Proc. Second International Conference on Knowledge Discovery and Data Mining, Portland, Oregon, E. Simoudis, J. Han and U. Fayyad, Eds., AAAI Press.

Povinelli, R. (2000). *Identifying Temporal Patterns for Characterization and Prediction of Financial Time Series Events*. In Proc. International Workshop on Temporal, Spatial and Spatio-Temporal Data Mining, TSDM2000, Lyon, France, J. F. Roddick and K. Hornsby, Eds., Springer. Lecture Notes in Artificial Intelligence. **2007**.

Srikant, R. and Agrawal, R. (1996). *Mining sequential patterns: generalisations and performance improvements.* In Proc. International Conference on Extending Database Technology, EDBT'96, Avignon, France, Peter M. G. Apers, M. Bouzeghoub and G. Gardarin, Eds., Springer-Verlag. Lecture Notes in Computer Science. **1057**: 3-17.

Tuzhilin, A. and Padmanabhan, B. (1996). *Pattern Discovery in Temporal Databases: A Temporal Logic Approach.* In Proc. 2nd International Conference on Knowledge Discovery and Data Mining, KDD'96, Portland, OR, E. Simoudis, J. Han and U. Fayyad, Eds., AAAI Press. 351-354.

Wade, T.D., Byrns, P.J., Steiner, J.F. and Bondy, J. (1994). Finding temporal patterns - a set based approach. *Artificial Intelligence in Medicine* **6**: 263-271.

Wang, K. (1997). Discovering patterns from large and dynamic sequential data. *Intelligent Information Systems* **8**: 8-33.

Wang, K. and Tan, J. (1996). *Incremental discovery of sequential patterns.* In Proc. ACM SIGMOD Workshop on Research Issues on Data Mining and Knowledge Discovery, Montreal, Canada.

Weiss, G.M. and Hirsh, H. (1998). *Learning to predict rare events in event sequences.* In Proc. Fourth International Conference on Knowledge Discovery and Data Mining (KDD'98), New York, NY, R. Agrawal, P. Stolorz and G. Piatetsky-Shapiro, Eds., AAAI Press, Menlo Park, CA. 359-363.

Wexelblat, A. (1996). *An environment for aiding information-browsing tasks.* In Proc. AAAI Spring Symposium on Acquisition, Learning and Demonstration: Automating Tasks for Users, Birmingham, England, AAAI Press.

Wijsen, J. and Meersman, R. (1997). *On the Complexity of Mining Temporal Trends.* In Proc. SIGMOD'97 Workshop on Research Issues on Data Mining and Knowledge Discovery, DMDK'97, Technical Report 97-07, Tucson, AZ, R. Ng, Ed. ACM Press. 77-84.

Zaki, M.J. (1998). *Efficient Enumeration of Frequent Sequences.* In Proc. Seventh International Conference on Information and Knowledge Management, Washington DC, 68-75.

Zaki, M.J., Lesh, N. and Ogihara, M. (1999). PlanMine: Predicting Plan Failures using Sequence Mining. *Artificial Intelligence Review, special issue on the Application of Data Mining.*

Zaki, M.J., Lesh, N. and Ogihara, M. (1998). *PlanMine: Sequence mining for plan failures.* In Proc. Fourth International Conference on Knowledge Discovery and Data Mining (KDD'98), New York, NY, R. Agrawal, P. Stolorz and G. Piatetsky-Shapiro, Eds., ACM Press. 369-373. A more detailed version appears in Artificial Intelligence Review, listed above.

2.4 Finding Similar Trends in Time Series

Agrawal, R., Psaila, G., Wimmers, E.L. and Zaöt, M. (1995). *Querying shapes of histories.* In Proc. Twenty-first International Conference on Very Large Databases (VLDB '95), Zurich, Switzerland, U. Dayal, P. M. D. Gray and S. Nishio, Eds., Morgan Kaufmann Publishers, Inc. San Francisco, USA. 502-514.

Berndt, D.J. and Clifford, J. (1995). Finding patterns in time series: a dynamic programming approach. In *Advances in Knowledge Discovery and Data Mining.* U. M. Fayyad, G. Piatetsky-Shapiro, P. Smyth and R. Uthurusamy, Eds. AAAI Press/ MIT Press. 229-248.

Clifford, J., Dhar, V. and Tuzhilin, A. (1995). Knowledge Discovery from Databases: The NYU Project. Technical Report IS-95-12. New York University New York.

Faloutsos, C., Ranganathan, M. and Manolopoulos, Y. (1994). *Fast subsequence matching in time-series databases.* In Proc. ACM SIGMOD Conference on the Management of Data, Minneapolis, USA, 419-429.

Guralnik, V. and Srivastava, J. (1999). *Event Detection from Time Series Data.* In Proc. Fifth International Conference on Knowledge Discovery and Data Mining, San Diego, CA, USA, S. Chaudhuri and D. Madigan, Eds., ACM Press. 33-42.

Han, J., Dong, G. and Yin, Y. (1999). *Efficient Mining of Partial Periodic Patterns in Time Series Database.* In Proc. Fifteenth International Conference on Data Engineering, Sydney, Australia, IEEE Computer Society. 106-115.

Keogh, E. and Smyth, P. (1997). *A probabilistic approach to fast pattern matching in time series databases.* In Proc. Third International Conference on Knowledge Discovery and Data Mining, Newport Beach, CA, USA, D. Heckerman, H. Mannila, D. Pregibon and R. Uthurusamy, Eds., AAAI Press, Menlo Park, California. 24-30.

Keogh, E.J. and Pazzani, M. (1999). *An Indexing Scheme for Fast Similarity Search in Large Time Series Databases.* In Proc. 11th International Conference on Scientific and Statistical Database Management, SSDBM'99, Cleveland, OH, IEEE Computer Society. 56-67.

Keogh, E.J. and Pazzani, M.J. (2000). *A Simple Dimensionality Reduction Technique for Fast Similarity Search in Large Time Series Databases.* In Proc. Knowledge Discovery and Data Mining, Current Issues and New Applications, 4th Pacific-Asia Conference, PAKDD 2000, Kyoto, Japan, T. Terano, H. Liu and A. Chen, Eds., Springer. Lecture Notes in Computer Science. **1805**: 122-133.

Povinelli, R. (2000). *Identifying Temporal Patterns for Characterization and Prediction of Financial Time Series Events.* In Proc. International Workshop on Temporal, Spatial and Spatio-Temporal Data Mining, TSDM2000, Lyon, France, J. F. Roddick and K. Hornsby, Eds., Springer. Lecture Notes in Artificial Intelligence. **2007**.

Tsumoto, S. (1999). *Rule Discovery in Large Time-Series Medical Databases.* In Proc. Principles of Data Mining and Knowledge Discovery, Third European

Conference, PKDD '99, Prague, Czech Republic, J. Zytkow and J. Rauch, Eds., Springer. Lecture Notes in Computer Science. **1704**: 23-31.

2.5 Discovery of Causal and Temporal Rules

Abraham, T. and Roddick, J.F. (1997). *Discovering meta-rules in mining temporal and spatio-temporal data.* In Proc. Eighth International Database Workshop, Data Mining, Data Warehousing and Client/Server Databases (IDW'97), Hong Kong, J. Fong, Ed. Springer-Verlag. 30-41.

Abraham, T. and Roddick, J.F. (1999). Incremental meta-mining from large temporal data sets. In *Advances in Database Technologies, Proc. First International Workshop on Data Warehousing and Data Mining, DWDM'98.* Y. Kambayashi, D. K. Lee, E.-P. Lim, M. Mohania and Y. Masunaga, Eds. Berlin. Springer-Verlag. Lecture Notes in Computer Science. **1552**: 41-54.

Agrawal, R. and Psaila, G. (1995). *Active Data Mining.* In Proc. First International Conference on Knowledge Discovery and Data Mining (KDD-95), Montreal, Quebec, Canada, U. Fayyad, M. and R. Uthurusamy, Eds., AAAI Press, Menlo Park, CA, USA. 3-8.

Bettini, C., Wang, X.S. and Jajodia, S. (1998). Mining Temporal Relationships with Multiple Granularities in Time Sequences. *Data Engineering Bulletin* **21**(1): 32-38.

Bettini, C., Wang, X.S. and Jajodia, S. (1996). *Testing Complex Temporal Relationships Involving Multiple Granularities and its Application to Data Mining.* In Proc. 15th ACM SIGACT-SIGMOD-SIGART Symposium on the Principles of Database Systems, Montreal, Canada, ACM Press. 68-78.

Blum, R.L. (1982). Discovery and Representation of Causal Relationships from a Large Time-Oriented Clinical Database: The RX Project. In *Lecture Notes in Medical Informatics.* Springer-Verlag. **19**.

Blum, R.L. (1982). Discovery, Confirmation and Interpretation of Causal Relationships from a Large Time-Oriented Clinical Database: The RX Project. *Computers and Biomedical Research* **15**(2): 164-187.

Chakrabarti, S., Sarawagi, S. and Dom, B. (1998). *Mining surprising patterns using temporal description length.* In Proc. Twenty-Fourth International Conference on Very Large databases VLDB'98, New York, NY, A. Gupta, O. Shmueli and J. Widom, Eds., Morgan Kaufmann. 606-617.

Chen, X. and Petrounias, I. (1998). *Language support for temporal data mining.* In Proc. Second European Symposium on Principles of Data Mining and Knowledge Discovery, PKDD'98, Nantes, France, J. M. Zytkow and M. Quafalou, Eds., Springer-Verlag, Berlin. Lecture Notes in Computer Science. **1510**: 282-290.

Chen, X., Petrounias, I. and Heathfield, H. (1998). *Discovering temporal association rules in temporal databases.* In Proc. International Workshop on Issues and Applications of Database Technology (IADT'98), 312-319.

Hamilton, H.J. and Randall, D.J. (1999). *Heuristic Selection of Aggregated Temporal Data for Knowledge Discovery.* In Proc. Multiple Approaches to Intelligent Systems, 12th International Conference on Industrial and Engineering Applications of Artificial Intelligence and Expert Systems, IEA/AIE-99,, Cairo, Egypt, I. Imam, Y. Kodratoff, A. El-Dessouki and M. Ali, Eds., Springer. Lecture Notes in Computer Science. **1611**: 714-723.

Hamilton, H.J. and Randall, D.J. (2000). *Data Mining with Calendar Attributes.* In Proc. International Workshop on Temporal, Spatial and Spatio-Temporal Data Mining, TSDM2000, Lyon, France, J. F. Roddick and K. Hornsby, Eds., Springer. Lecture Notes in Artificial Intelligence. **2007**.

Hickey, R. and Black, M.M. (2000). *Refined Time Stamps for Concept Drift Detection During Mining for Classification Rules.* In Proc. International Workshop on Temporal, Spatial and Spatio-Temporal Data Mining, TSDM2000, Lyon, France, J. F. Roddick and K. Hornsby, Eds., Springer. Lecture Notes in Artificial Intelligence. **2007**.

Imam, I.F. (1994). *An experimental study of discovery in large temporal databases.* In Proc. Seventh International Conference on Industrial and Engineering Applications of Artificial Intelligence and Expert Systems, IEA/AIE '94., 171-180.

Keogh, E. and Pazzani, M. (1999). *Scaling up Dynamic Time Warping to Massive Datasets.* In Proc. 3rd European Conference on Principles and Practice of Knowledge Discovery in Databases (PKDD'99), Prague, Czech Republic, J. M. Zytkow and J. Rauch, Eds., Springer. Lecture Notes in Artificial Intelligence. **1704**: 1-11.

Long, J.M., Irani, E.A. and Slagle, J.R. (1991). Automating the Discovery of Causal Relationships in a Medical Records Database. In *Knowledge discovery in databases.* G. Piatetsky-Shapiro and W. J. Frawley, Eds. AAAI Press/MIT Press. 465-476.

Rainsford, C.P. and Roddick, J.F. (1997). *An attribute-oriented induction of rules from temporal interval data.* In Proc. Eighth International Database Workshop, Data Mining, Data Warehousing and Client/Server Databases (IDW'97), Hong Kong, J. Fong, Ed. Springer Verlag. 108-118.

Rainsford, C.P. and Roddick, J.F. (2000). *Temporal Interval Logic in Data Mining.* In Proc. Sixth Pacific Rim International Conference on Artificial Intelligence, PRICAI2000, Melbourne, R. Mizoguchi and J. Slaney, Eds., Springer. Lecture Notes in Artificial Intelligence. **1886**: 798.

Rainsford, C.P. and Roddick, J.F. (2000). *Visualisation of Temporal Interval Association Rules.* In Proc. 2nd International Conference on Intelligent Data Engineering and Automated Learning, (IDEAL 2000), Shatin, N.T., Hong Kong.

Saraee, M.H. and Theodoulidis, B. (1995). *Knowledge discovery in temporal databases.* In Proc. IEE Colloquium on 'Knowledge Discovery in Databases', IEE, London. 1-4.

Sasisekharan, R., Seshadri, V. and Weiss, S.M. (1996). Data Mining and Forecasting in Large-Scale Telecommunication Networks. *IEEE Expert* **11**(1): 37-43.

2.6   Spatial Data Mining Publications

Andrienko, G.L. and Andrienko, N.V. (1999). *Knowledge-Based Visualization to Support Spatial Data Mining.* In Proc. Third International Symposium on Advances in Intelligent Data Analysis, IDA-99, Amsterdam, The Netherlands, D. J. Hand, J. N. Kok and M. R. Berthold, Eds., Springer. Lecture Notes in Computer Science. **1642**: 149-160.

Ankerst, M., Kastenmller, G., Kriegel, H.-P. and Seidl, T. (1999). *3D Shape Histograms for Similarity Search and Classification in Spatial Databases.* In Proc. Advances in Spatial Databases, 6th International Symposium, SSD'99, Hong Kong, China, R. H. Güting, D. Papadias and F. H. Lochovsky, Eds., Springer. **1651**: 207-228.

Bell, D.A., Anand, S.S. and Shapcott, C.M. (1994). *Data Mining in Spatial Databases.* In Proc. International Workshop on Spatio-Temporal Databases, Benicassim, Spain.

Bigolin, N.M. and Marsala, C. (1998). *Fuzzy Spatial OQL for Fuzzy Knowledge Discovery in Databases.* In Proc. Second European Symposium on the Principles of Data Mining and Knowledge Discovery, PKDD'98, Nantes, France, J. M. Zytkow and M. Quafalou, Eds., Springer-Verlag, Berlin. Lecture Notes in Computer Science. **1510**: 246-254.

Chawla, S., Shekhar, S., Wu, W. and Ozesmi, U. (2000). *Extending Data Mining for Spatial Applications: A Case Study in Predicting Nest Locations.* In Proc. ACM SIGMOD Workshop on Research Issues in Data Mining and Knowledge Discovery, Texas, USA, D. Gunopulos and R. Rastogi, Eds., ACM. 70-77.

Ester, M., Frommelt, A., Kriegel, H.-P. and Sander, J. (2000). Spatial Data Mining: Database Primitives, Algorithms and Efficient DBMS Support. *Data Mining and Knowledge Discovery* 4(2/3): 193-216.

Ester, M., Frommelt, A., Kriegel, H.P. and Sander, J. (1998). *Algorithms for characterization and trend detection in spatial databases.* In Proc. Fourth International Conference on Knowledge Discovery and Data Mining., New York, NY, R. Agrawal, P. Stolorz and G. Piatetsky-Shapiro, Eds., AAAI Press, Menlo Park. 44-50.

Ester, M., Gundlach, S., Kriegel, H.-P. and Sander, J. (1999). *Database Primitives for Spatial Data Mining.* In Proc. International Conference on Databases in Office, Engineering and Science, BTW'99, Freiberg, Germany, 137-150.

Ester, M., Kriegel, H.-P. and Sander, J. (1999). *Knowledge Discovery in Spatial Databases.* In Proc. 23rd German Conference on Artificial Intelligence, KI'99, Bonn, Germany, Springer. Lecture Notes in Computer Science. **1701**: 61-74.

Ester, M., Kriegel, H.-P. and Sander, J. (1997). *Spatial Data Mining: A Database Approach.* In Proc. Fifth Symposium on Large Spatial Databases (SSD'97), Berlin, Germany, M. Scholl and A. Voisard, Eds., Springer. Lecture Notes in Computer Science. **1262**: 48-66.

Ester, M., Kriegel, H.-P., Sander, J. and Xu, X. (1996). *A Density-Based Algorithm for Discovering Clusters in Large Spatial Databases with Noise*. In Proc. Second International Conference on Knowledge Discovery and Data Mining, Portland, Oregon, E. Simoudis, J. Han and U. Fayyad, Eds., AAAI Press.

Ester, M., Kriegel, H.-P. and Xu, X. (1995). *A Database Interface for Clustering in Large Spatial Databases*. In Proc. First International Conference on Knowledge Discovery and Data Mining, KDD'95, Montreal, Canada, AAAI Press. 94-99.

Ester, M., Kriegel, H.-P. and Xu, X. (1995). *Knowledge discovery in large spatial databases: focusing techniques for efficient class identification*. In Proc. Advances in Spatial Databases, 4th International Symposium, SSD'95, Portland, ME, M. Egenhofer and J. Herring, Eds., Springer. Lecture Notes in Computer Science. **951**: 67-82.

Ester, M. and Wittmann, R. (1998). *Incremental Generalization for Mining in a Data Warehousing Environment*. In Proc. Sixth International Conference on Extending Database Technology, Valencia, Spain, H.-J. Schek, F. Saltor and I. Ramos, Eds., Springer. Lecture Notes in Computer Science. **1377**: 135-149.

Estivill-Castro, V. and Houle, M. (2000). *Fast Randomized Algorithms for Robust Estimation of Location*. In Proc. International Workshop on Temporal, Spatial and Spatio-Temporal Data Mining, TSDM2000, Lyon, France, J. F. Roddick and K. Hornsby, Eds., Springer. Lecture Notes in Artificial Intelligence. **2007**.

Estivill-Castro, V. and Lee, I. (2000). *AUTOCLUST+: Automatic Clustering of Point-Data Sets in the Presence of Obstacles*. In Proc. International Workshop on Temporal, Spatial and Spatio-Temporal Data Mining, TSDM2000, Lyon, France, J. F. Roddick and K. Hornsby, Eds., Springer. Lecture Notes in Artificial Intelligence. **2007**.

Estivill-Castro, V. and Murray, A.T. (1998). *Discovering associations in spatial data-an efficient mediod based approach*. In Proc. Second Pacific-Asia Conference on Research and Development in Knowledge Discovery and Data Mining, PAKDD-98, Springer-Verlag, Berlin. 110-121.

Han, J., Koperski, K. and Stefanovic, N. (1997). *GeoMiner: A System Prototype for Spatial Data Mining*. In Proc. ACM SIGMOD Conference on the Management of Data, Tucson, Arizona, USA, J. Peckham, Ed. ACM Press. 553-556.

Han, J., Stefanovic, N. and Koperski, K. (1998). *Selective Materialization: An Efficient Method for Spatial Data Cube Construction*. In Proc. Research and Development in Knowledge Discovery and Data Mining, Second Pacific-Asia Conference, PAKDD'98, Melbourne, Australia, X. Wu, K. Ramamohanarao and K. Korb, Eds., 144-158.

Kang, I.-S., Kim, T.-W. and Li, K.-J. (1997). *A Spatial Data Mining Method by Delaunay Triangulation*. In Proc. Fifth ACM Workshop on Geographic Information Systems, GIS'97, Las Vegas, Nevada, 35-39.

Knorr, E.M. and Ng, R.T. (1996). *Extraction of Spatial Proximity Patterns by Concept Generalization.* In Proc. Second International Conference on Knowledge Discovery and Data Mining, Portland, Oregon, E. Simoudis, J. Han and U. Fayyad, Eds., AAAI Press. 347-350.

Knorr, E.M. and Ng, R.T. (1996). Finding aggregate proximity relationships and commonalities in spatial data mining. *IEEE Transactions on Knowledge and Data Engineering* **8**(6): 884-897.

Knorr, E.M., Ng, R.T. and Shilvock, D.L. (1997). Finding boundary shape matching relationships in spatial data. In *Advances in Spatial Databases - Proc. 5th International Symposium, SSD '97.* Springer-Verlag, Berlin. 29-46.

Koperski, K., Adhikary, J. and Han, J. (1996). *Knowledge Discovery in Spatial Databases: Progress and Challenges.* In Proc. ACM SIGMOD Workshop on Research Issues on Data Mining and Knowledge Discovery, Montreal, Canada, 55-70.

Koperski, K. and Han, J. (1995). *Discovery of Spatial Association Rules in Geographic Information Databases.* In Proc. Fourth International Symposium on Large Spatial Databases, Maine, 47-66.

Lin, X., Zhou, X. and Liu, C. (1999). *Efficiently Matching Proximity Relationships in Spatial Databases.* In Proc. Advances in Spatial Databases, 6th International Symposium, SSD'99, Hong Kong, China, R. H. Güting, D. Papadias and F. H. Lochovsky, Eds., Springer. Lecture Notes in Computer Science. **1651**: 188-206.

Lu, W., Han, J. and Ooi, B.C. (1993). *Discovery of General Knowledge in Large Spatial Databases.* In Proc. 1993 Far East Workshop on GIS (IEGIS 93), Singapore, 275-289.

Miller, H. and Han, J. (1999). Discovering Geographic Knowledge in Data-Rich Environments. Report of a Specialist Meeting held under the auspices of the Varenius Project Kirkland, WA.

Ng, R.T. (1996). *Spatial Data Mining: Discovering Knowledge of Clusters from Maps.* In Proc. ACM SIGMOD Workshop on Research Issues on Data Mining and Knowledge Discovery, Montreal, Canada.

Ng, R.T. and Han, J. (1994). *Efficient and effective clustering methods for spatial data mining.* In Proc. Twentieth International Conference on Very Large Data Bases, Santiago, Chile, J. B. Bocca, M. Jarke and C. Zaniolo, Eds., Morgan Kaufmann. 144-155.

Ng, R.T. and Yu, Y. (1997). *Discovering Strong, Common and Discriminating Characteristics of Clusters from Thematic Maps.* In Proc. 11th Annual Symposium on Geographic Information Systems, 392-394.

Popelinsky, L. (1998). *Knowledge discovery in spatial data by means of ILP.* In Proc. Second European Symposium on the Principles of Data Mining and Knowledge Discovery, PKDD'98, Nantes, France, J. M. Zytkow and M. Quafalou, Eds., Springer-Verlag, Berlin. Lecture Notes in Computer Science. **1510**: 185-193.

Sander, J., Ester, M., Kriegel, H.-P. and Xu, X. (1998). Density-Based Clustering in Spatial Databases: The Algorithm GDBSCAN and Its Applications. *Data Mining and Knowledge Discovery* **2**(2): 169-194.

Shek, E.C., Muntz, R.R., Mesrobian, E. and Ng, K. (1996). *Scalable Exploratory Data Mining of Distributed Geoscientific Data.* In Proc. Second International Conference on Knowledge Discovery and Data Mining, Portland, Oregon, E. Simoudis, J. Han and U. Fayyad, Eds., AAAI Press. 32-37.

Son, E.-J., Kang, I.-S., Kim, T.-W. and Li, K.-J. (1998). *A Spatial Data Mining Method by Clustering Analysis.* In Proc. Sixth International Symposium on Advances in Geographic Information Systems, GIS'98, Washington, DC, USA, ACM. 157-158.

Wang, W., Yang, J. and Muntz, R. (1999). *STING+: An approach to active spatial data mining.* In Proc. Fifteenth International Conference on Data Engineering, Sydney, Australia, IEEE Computer Society. 116-125.

Wang, W., Yang, J. and Muntz, R.R. (2000). An Approach to Active Spatial Data Mining Based on Statistical Information. *IEEE Transactions on Knowledge and Data Engineering* **12**(5): 715-728.

Wang, W., Yang, J. and Muntz, R.R. (1997). *STING: A Statistical Information Grid Approach to Spatial Data Mining.* In Proc. Twenty-Third International Conference on Very Large Data Bases, Athens, Greece, M. Jarke, *et al*, Eds., Morgan Kaufmann. 186-195.

Xu, X., Ester, M., Kriegel, H.-P. and Sander, J. (1998). *A Distribution-Based Clustering Algorithm for Mining in Large Spatial Databases.* In Proc. Fourteenth International Conference on Data Engineering, ICDE'98, Orlando, Florida, USA, IEEE Computer Society. 324-331.

Yang, R., Yang, K.-S., Kafatos, M. and Wang, X.S. (2000). *Value Range Queries on Earth Science Data via Histogram Clustering.* In Proc. International Workshop on Temporal, Spatial and Spatio-Temporal Data Mining, TSDM2000, Lyon, France, J. F. Roddick and K. Hornsby, Eds., Springer. Lecture Notes in Artificial Intelligence. **2007**.

Zeitouni, K., Yeh, L. and Aufaure, M.-A. (2000). *Join Indices are a Tool for Spatial Data Mining.* In Proc. International Workshop on Temporal, Spatial and Spatio-Temporal Data Mining, TSDM2000, Lyon, France, J. F. Roddick and K. Hornsby, Eds., Springer. Lecture Notes in Artificial Intelligence. **2007**.

Zhang, B., Hsu, M. and Dayal, U. (2000). *K-Harmonic Means: A Spatial Clustering Algorithm with Boosting.* In Proc. International Workshop on Temporal, Spatial and Spatio-Temporal Data Mining, TSDM2000, Lyon, France, J. F. Roddick and K. Hornsby, Eds., Springer. Lecture Notes in Artificial Intelligence. **2007**.

Zhou, X., Truffet, D. and Han, J. (1999). *Efficient Polygon Amalgamation Methods for Spatial OLAP and Spatial Data Mining.* In Proc. 6th International Symposium on Spatial Databases (SSD'99), Hong Kong, R. H. Güting, D. Papadias and F. H. Lochovsky, Eds., Springer. Lecture Notes in Computer Science. **1651**: 167-187.

2.7   Spatial and Spatio-temporal Clustering Techniques

Ester, M., Kriegel, H.-P., Sander, J. and Xu, X. (1996). *A Density-Based Algorithm for Discovering Clusters in Large Spatial Databases with Noise*. In Proc. Second International Conference on Knowledge Discovery and Data Mining, Portland, Oregon, E. Simoudis, J. Han and U. Fayyad, Eds., AAAI Press.

Ester, M., Kriegel, H.-P. and Xu, X. (1995). *A Database Interface for Clustering in Large Spatial Databases*. In Proc. First International Conference on Knowledge Discovery and Data Mining, KDD'95, Montreal, Canada, AAAI Press. 94-99.

Estivill-Castro, V. and Lee, I. (2000). *AUTOCLUST+: Automatic Clustering of Point-Data Sets in the Presence of Obstacles*. In Proc. International Workshop on Temporal, Spatial and Spatio-Temporal Data Mining, TSDM2000, Lyon, France, J. F. Roddick and K. Hornsby, Eds., Springer. Lecture Notes in Artificial Intelligence. **2007**.

Keogh, E. and Pazzani, M. (1998). *An enhanced representation of time series which allows fast and accurate classification, clustering and relevance feedback*. In Proc. Fourth International Conference on Knowledge Discovery and Data Mining (KDD'98), New York City, NY, R. Agrawal, P. Stolorz and G. Piatetsky-Shapiro, Eds., ACM Press. 239-241.

Ketterlin, A. (1997). *Clustering Sequences of Complex Objects*. In Proc. Third International Conference on Knowledge Discovery and Data Mining, Newport Beach, CA, USA, D. Heckerman, H. Mannila, D. Pregibon and R. Uthurusamy, Eds., AAAI Press, Menlo Park, California. 215-218.

Ng, R.T. (1996). *Spatial Data Mining: Discovering Knowledge of Clusters from Maps*. In Proc. ACM SIGMOD Workshop on Research Issues on Data Mining and Knowledge Discovery, Montreal, Canada.

Ng, R.T. and Han, J. (1994). *Efficient and effective clustering methods for spatial data mining*. In Proc. Twentieth International Conference on Very Large Data Bases, Santiago, Chile, J. B. Bocca, M. Jarke and C. Zaniolo, Eds., Morgan Kaufmann. 144-155.

Oates, T. (1999). *Identifying Distinctive Subsequences in Multivariate Time Series by Clustering*. In Proc. Fifth International Conference on Knowledge Discovery and Data Mining, San Diego, CA, USA, S. Chaudhuri and D. Madigan, Eds., ACM Press. 322-326.

Sander, J., Ester, M., Kriegel, H.-P. and Xu, X. (1998). Density-Based Clustering in Spatial Databases: The Algorithm GDBSCAN and Its Applications. *Data Mining and Knowledge Discovery* **2**(2): 169-194.

Son, E.-J., Kang, I.-S., Kim, T.-W. and Li, K.-J. (1998). *A Spatial Data Mining Method by Clustering Analysis*. In Proc. Sixth International Symposium on Advances in Geographic Information Systems, GIS'98, Washington, DC, USA, ACM. 157-158.

Xu, X., Ester, M., Kriegel, H.-P. and Sander, J. (1998). *A Distribution-Based Clustering Algorithm for Mining in Large Spatial Databases*. In Proc. Fourteenth International Conference on Data Engineering, ICDE'98, Orlando, Florida, USA, IEEE Computer Society. 324-331.

Yang, R., Yang, K.-S., Kafatos, M. and Wang, X.S. (2000). *Value Range Queries on Earth Science Data via Histogram Clustering.* In Proc. International Workshop on Temporal, Spatial and Spatio-Temporal Data Mining, TSDM2000, Lyon, France, J. F. Roddick and K. Hornsby, Eds., Springer. Lecture Notes in Artificial Intelligence. **2007**.

Zhang, B., Hsu, M. and Dayal, U. (2000). *K-Harmonic Means: A Spatial Clustering Algorithm with Boosting.* In Proc. International Workshop on Temporal, Spatial and Spatio-Temporal Data Mining, TSDM2000, Lyon, France, J. F. Roddick and K. Hornsby, Eds., Springer. Lecture Notes in Artificial Intelligence. **2007**.

Zhang, T., Ramakrishnan, R. and Livny, M. (1996). *BIRCH: An Efficient Clustering Method for Very Large Databases.* In Proc. ACM SIGMOD Workshop on Research Issues on Data Mining and Knowledge Discovery, Montreal, Canada, 103-114.

2.8   Spatio-temporal Data Mining

Abraham, T. and Roddick, J.F. (1997). *Discovering meta-rules in mining temporal and spatio-temporal data.* In Proc. Eighth International Database Workshop, Data Mining, Data Warehousing and Client/Server Databases (IDW'97), Hong Kong, J. Fong, Ed. Springer-Verlag. 30-41.

Bittner, T. (2000). *Rough Sets in Spatio-Temporal Data Mining.* In Proc. International Workshop on Temporal, Spatial and Spatio-Temporal Data Mining, TSDM2000, Lyon, France, J. F. Roddick and K. Hornsby, Eds., Springer. Lecture Notes in Artificial Intelligence. **2007**.

Klösgen, W. (1995). *Deviation and association patterns for subgroup mining in temporal, spatial, and textual data bases.* In Proc. First International Conference on Rough Sets and Current Trends in Computing, RSCTC'98, Springer-Verlag, Berlin,. 1-18.

Klösgen, W. (1998). *Subgroup mining in temporal, spatial and textual databases.* In Proc. International Symposium on Digital Media Information Base, World Scientific, Singapore. 246-261.

Mesrobian, E., *et al.* (1996). Mining geophysical data for knowledge. *IEEE Expert* **11**(5): 34-44.

Mesrobian, E., *et al.* (1995). *Exploratory data mining and analysis using CONQUEST.* In Proc. IEEE Pacific Rim Conference on Communications, Computers and Signal Processing, IEEE, New York. 281-286.

Peuquet, D. and Wentz, E. (1994). *An Approach for Time-Based Analysis of Spatiotemporal Data.* In Proc. Sixth International Symposium on Spatial Data Handling, Edinburgh, Scotland, R. G. Healy, Ed. Taylor and Francis. Advances in GIS Research.

Roddick, J.F. and Hornsby, K., Eds. (2001). *Temporal, Spatial and Spatio-Temporal Data Mining. Proc. First International Workshop.* Lecture Notes in Computer Science. Berlin-Heidelberg, Springer.

Stolorz, P. and Dean, C. (1996). *Quakefinder: A Scalable Data Mining System for Detecting Earthquakes from Space*. In Proc. Second International Conference on Knowledge Discovery and Data Mining (KDD96), Portland, Oregon, E. Simoudis, J. Han and U. Fayyad, Eds., AAAI Press, Menlo Park, CA, USA. 208-213.

Stolorz, P., *et al.* (1995). *Fast Spatio-Temporal Data Mining of Large Geophysical Sets*. In Proc. First International Conference on Knowledge Discovery and Data Mining, Montreal, Canada, AAAI Press. 300-305.

Valdes-Perez, R.E. (1998). Systematic Detection of Subtle Spatio-Temporal Patterns in Time-Lapse Imaging. I. Mitosis. *Bioimaging* **4**(4): 232-242.

Valdes-Perez, R.E. and Stone, C.A. (1998). Systematic Detection of Subtle Spatio-Temporal Patterns in Time-Lapse Imaging II. Particle Migrations. *Bioimaging* **6**(2): 71-78.

Wijsen, J. and Ng, R.T. (1999). *Temporal Dependencies Generalized for Spatial and Other Dimensions*. In Proc. International Workshop on Spatio-Temporal Database Management, Edinburgh, Scotland, Springer. Lecture Notes in Computer Science. **1678**: 189-203.

2.9 Theses, Surveys, Books, and Previous Bibliographies

Abraham, T. (1999). Knowledge Discovery in Spatio-Temporal Databases. PhD Thesis. University of South Australia.

Abraham, T. and Roddick, J.F. (1998). Opportunities for knowledge discovery in spatio-temporal information systems. *Australian Journal of Information Systems* **5**(2): 3-12.

Koperski, K. (1999). Progressive Refinement Approach to Spatial Data Mining. Ph.D. Thesis. Simon Fraser University.

Miller, H. and Han, J., Eds. (2001). *Geographic Data Mining and Knowledge Discovery*. Research Monographs in Geographic Information Systems, Taylor and Francis.

Rainsford, C.P. (1999). Accommodating Temporal Semantics in Data Mining and Knowledge Discovery. PhD Thesis. University of South Australia.

Roddick, J.F. and Spiliopoulou, M. (1999). A Bibliography of Temporal, Spatial and Spatio-Temporal Data Mining Research. *SIGKDD Explorations* **1**(1): 34-38.

Roddick, J.F. and Spiliopoulou, M. (2001). A Survey of Temporal Knowledge Discovery Paradigms and Methods. *IEEE Transactions on Knowledge and Data Engineering*.

Weigend, A.S. and Gershenfeld, N.A., Eds. (1993). *Time Series Prediction: Forecasting the Future and Understanding the Past*. Proc. NATO Advanced Research Workshop on Comparative Time Series Analysis. Santa Fe, New Mexico, Addison-Wesley.

3   Conferences

This section lists a number of the conferences that have produced relevant publications in the past and thus could be worth monitoring. Some are run periodically while some have been single one-off workshops. In general, most have not focussed specifically on temporal, spatial or spatio-temporal data mining but rather have included sessions or individual papers on the topic. Some forums, while producing papers that have been included above, are unlikely to do so in the future and have therefore been omitted.

- ACM SIGKDD Conference on Knowledge Discovery and Data Mining.
- ACM SIGMOD Conference on the Management of Data.
- ACM SIGMOD Workshops on Research Issues on Data Mining and Knowledge Discovery.
- ACM Workshop on Geographic Information Systems.
- European Conference on Principles and Practice of Knowledge Discovery in Databases (PKDD Series).
- IEE Colloquium on 'Knowledge Discovery in Databases'.
- International Conference on Data Engineering (ICDE Series).
- International Conference on Database and Expert Systems Applications, (DEXA series).
- International Conference on Extending Database Technology, (EDBT series).
- International Conference on Information and Knowledge Management (CIKM series).
- International Conference on Very Large Data Bases (VLDB series).
- International Database Workshop series.
- International Symposium on Advances in Geographic Information Systems.
- International Symposium on Spatial Data Handling.
- International Symposium on Spatial and Temporal Databases (SSTD), formerly the Symposium on Spatial Databases (SSD Series).
- International Workshop on Spatio-Temporal Databases.
- International Workshop on Temporal, Spatial and Spatio-Temporal Data Mining, (TSDM2000, possibly to become a series).
- Pacific-Asia Conference on Research and Development in Knowledge Discovery and Data Mining, (PAKDD series).
- SIAM International Conference on Data Mining (SDM series).

The bibliography is being kept up to date at
    http://www.cs.flinders.edu.au/People/John_Roddick/STDMPapers/.
Additions and comments welcome.

Author Index

Lecture Notes in Artificial Intelligence (LNAI)

Lecture Notes in Computer Science

GPSR Compliance

The European Union's (EU) General Product Safety Regulation (GPSR)
is a set of rules that requires consumer products to be safe and our
obligations to ensure this.

If you have any concerns about our products, you can contact us on
ProductSafety@springernature.com

In case Publisher is established outside the EU, the EU authorized
representative is:

Springer Nature Customer Service Center GmbH
Europaplatz 3
69115 Heidelberg, Germany

Batch number: 09625304

Printed by Printforce, the Netherlands